PRAISE FOR KATY BOWMAN

"When it comes to your body, Katy Bo[...] friend you've been looking for your whole life. Having Katy in your life is like having a superpower."

—**Kelly Starrett**, DPT, *New York Times* bestselling author of *Becoming a Supple Leopard*, *Deskbound*, and *Built to Move*

"Katy Bowman has been reminding us for years that, just as food is medicine, a nutritious movement diet is essential for optimal health and well-being."

—**Mark Hyman, MD**, #1 *New York Times* bestselling author of *Young Forever*, *The Pegan Diet*, and *The Blood Sugar Solution*

"Katy Bowman's work always fires both my imagination and my admiration. Refined through her practice, her parenting (so important), and her unwavering use of herself as an experiment, Katy's whole body of work is pointing toward an updated physical education for the coming challenges."

—**Tom Myers**, author of *Anatomy Trains*

"Movement is so much more than exercise. Our posture when we sit and stand, how we walk, bend over to pick something up, and even the micro-movements we make when at rest are vital to our health and well-being. Katy Bowman helps us to reshape our everyday movement and find comfort and ease in our bodies. I consider (her) essential reading for all humans!"

—**Chris Kresser**, M.S., L.Ac., *New York Times* bestselling author of *Paleo Cure*

"Katy Bowman helps you take better care of yourself by helping you understand how your body works and the simple adjustments it needs to be its best. Her advice is easy to understand, practical, and effective, whether you're in a wheelchair or an elite athlete or anywhere in between. Perhaps best of all, she helps you see opportunities to get the movements your body needs everywhere you go."

—**Kate Hanley**, author of *How to Be a Better Person* and host of the *How to Be a Better Person* podcast

"Katy Bowman invites you to easily meet and greet body areas that may have departed from inclusive movement over the course of your life, while making you feel at home inside yourself."

—**Jill Miller**, author of *Body by Breath* and *The Roll Model*, Co-Founder of Tune Up Fitness Worldwide

My Perfect Movement Plan

My Perfect Movement Plan

THE MOVE YOUR DNA ALL DAY WORKBOOK

Katy Bowman, M.S.

UPHILL BOOKS
MOVEMENT MATTERS

Printed in the United States of America
First Edition, First Printing, 2024
ISBN-13: 9781943370269
Library of Congress Control Number: 2024935978

Uphill Books: uphill-books.com

Editor: Penelope Jackson
Science Editor: Andrea Graves
Text & Cover Design by: Agnes Koller, figdesign.ca
Proofreader: Kate Kennedy
Indexer: Michael Curran

Image/Photo Credits: p.89 Jillian Nicol; p. 91 Mahina Hawley; p.95 Agnes Koller, figdesign.ca; p. 164 Cecilia Ortiz; p. 207 Jillian Nicol; p. 210 Shelah M. Wilgus; Author photo by Mahina Hawley

The information in this book should not be used for diagnosis or treatment, or as a substitute for professional medical care. Please consult with your health care provider prior to attempting any treatment on yourself or another individual.

Names:	Bowman, Katy, author.
Title:	My perfect movement plan : the move your DNA all day workbook / Katy Bowman, M.S.
Description:	First edition.	[Sequim, Washington] : Uphill Books, [2024]	Includes bibliographical references and index.
Identifiers:	ISBN: 9781943370269 (paperback)	9781943370276 (ebook)	LCCN: 2024935978
Subjects:	LCSH: Movement education.	Physical fitness.	Exercise.	Human mechanics.	Kinesiology.	BISAC: HEALTH & FITNESS / Exercise / General.	SCIENCE / Life Sciences / Human Anatomy & Physiology.	HEALTH & FITNESS / Exercise / Stretching.
Classification:	LCC: GV452 .B69 2024	DDC: 372.86--dc23

For Debbie B., who always shines
on her people and place, most often
through her language of love: labor

ALSO BY KATY BOWMAN

Contents

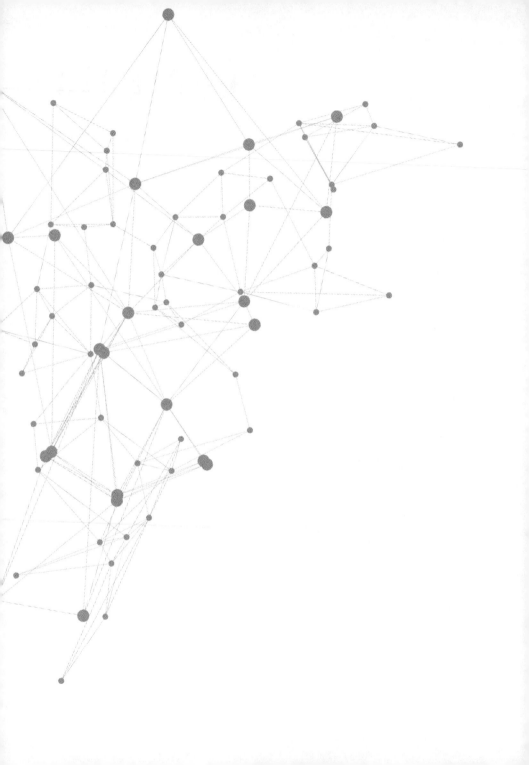

A Note on Writing in Books

As you embark on creating your own perfect movement plan, you may wonder why this is a book and not an app. Wouldn't it be easier to tap out answers on a screen and have an algorithm spit out the movements we need?

Maybe—but it would be less effective. Research shows there are benefits to writing things down. For example, MRI studies have demonstrated that writing on paper activates areas of the brain differently from how writing and interacting with a screen does. There is more spatial information in physical writing; unlike the uniform type and disappearance of scrolling screens that come with digital writing, physical writing is a richer spatial experience, which helps our brains connect better to the material. Handwriting is also a more active process!

And, while the end goal is to create a map to finding the movements you need, the way you get there is through deeply questioning and thinking about yourself and learning to see yourself in a new way. The physical act of writing these answers down is part of why journaling is so helpful—it helps the mind create order while also doing something that requires some physical coordination.

So please, write away! Don't just skim along answering the questions in your head. Take the time and care to deeply reflect and to write out your answers longhand. This is a workbook for a reason. It's incomplete until you complete it. And while you're self-assessing, answering questions, and tracking your movements, feel free to also doodle in the margins, make notes and reminders, and record your discoveries.

Unless of course...*this isn't your copy.*

Did you get this from a library or a friend? Do you want to share it with your partner or your whole family? Do you want to use the book again in six months to track and refine your movement plan? Are you reading this as an e-book?

I'VE GOT YOU. Find a PDF of the activity pages only, which you can download for free as many times as you need, at nutritiousmovement.com/MPMPresources.

There. Now you have no excuses. Let's get to work!

Introduction: Move More or Move Better?

I've studied and taught movement for over twenty-five years. When people learn that about me, they always ask something like "What's the best exercise?" or "What three exercises should I be doing daily?" or "What exercises can fix my _____ problem?"

The answer is never simple, because it depends on the answers to a lot of other questions, such as:

- » What types of activities are you doing now, and for how long?
- » What would you most like to be able to do in the future?
- » What injuries have you had?
- » Which of your parts are already getting cranky, even without injury?

» What stage of life are you in?

» How is your environment currently set up for movement?

Everyone's perfect movement plan is different based on all of these questions and more; there will never be one plan to move them all. That's where *My Perfect Movement Plan* comes in.

This book is all about discovering the movement your body needs. Not any body, but your body, specifically. I've written most of the book, and once you've written the rest of it, you'll have a plan, your best plan, for nourishing *your* body with movement in a way that works for you and your lifestyle.

Movement is simple: it happens anytime you change the position of your body in some way. But answering the questions above takes work. The answers require your frank assessment of what you want to do with your body in the future, what you have done with your body before, and how you're able to move today.

The answers also require a lot more nuance than we're used to applying to movement. In general, we have a very simple understanding of movement as something a whole body does, but in fact each of our parts—and we've got a lot of parts—need their own movement. Certain movements, like swimming, for example, can be great for the whole body and also get the arms and legs moving, but that activity doesn't do much to maintain bone strength (bones need to feel heavy loads to stay dense, and the buoyant force of water makes us light). Likewise, if you sit in front of a computer for multiple hours each day and then hop onto your bike for your dose of daily movement, your hips never really spend much time out of a seat and are barely ever moved. Maybe your job has you walking everywhere, but your arms almost never reach overhead and your neck and shoulders get less and less mobile each year.

DO YOU NEED TO MOVE MORE, MOVE BETTER, OR BOTH?

When it comes to movement, we often think in terms of getting *enough*. But movement is more complicated than that. Movement works similarly to how food works in the body. The movement you put "in"to your body affects how its tissues and cells work, and you're never just putting movement into your whole body—you're putting movement into the specific parts that are doing the work.

Let's think about food diets for a second. These days, most people are aware that the effects of their diet are not based simply on getting enough food. The ratio of carbohydrates, proteins, dietary fiber, fats, vitamins, and minerals matters tremendously. Even the timing of meals can factor into the way diet impacts the body's function.

Just as we talk about our food diet in terms of calories, macronutrients, and micronutrients, we can—and should!—think of movement in the same way. When you move, no matter which parts are doing it, you're generating "movement calories"—and the type of movement you're doing (i.e., the activity and which parts are working) contains particular "movement nutrients."

Eating a balanced food diet means we get our calories from the three macronutrient categories—fat, protein, and carbohydrate. Similarly, a balanced movement diet needs to include movements from a range of categories: endurance, strength, mobility, and balance. Instead of looking solely at *how much*, in steps or minutes, we move each day (think "movement calories"), we also need to make sure we're using movement in a variety of ways and distributing that movement all over our body.

Movement is an essential input for the body, and across the board there's agreement that we must move our bodies for health and longevity. But there are some issues in this simplicity. Laborers move their bodies all day long, yet many leave their work due to injury and pain—or worse, must keep working despite the aches. All that movement isn't making their bodies feel good. Athletes can train hard and be fantastically fit while they're performing but end up not being able to move well later in life. Shouldn't moving now help us move better in the future? And what about those who desire to move more but can't because it hurts when they move? The simplistic directive to "Just move more!" doesn't work for many people—it often requires elaboration.

I began using a "movement nutrient" framework fifteen years ago when looking for an efficient way to explain movement in the above variety of experiences and more. I even named my movement company Nutritious Movement because the calorie/macronutrient/micronutrient framework helps people understand the breadth of movement our bodies really need. It helps us understand which movements we aren't getting enough of and which we might be getting too much of. I'll reference this framework throughout this book, and I hope that by the end you will see movement in an entirely different light.

In the most general sense, there are three broad classifications of food diets: those balanced in terms of calories and nutrients (the right amount of food containing the correct quantities of macro- and micronutrients); those high in calories but low in nutrients (you're getting enough energy from the calories, but you're still missing essential nutrients); and those too low in both calories and necessary nutrients. "Movement diets" work in the same way. If "movement calories" are the total units (minutes) of movement you're getting and "movement nutrients" are the different shapes your body flows through to create that movement, which below sounds the closest to your movement diet?

High-volume, nutrient-dense movement/a well-balanced movement diet:	You move your body a large portion of the day in a way that nourishes *all* the different body parts and tissue-types, and also develops (or maintains) the movement skills necessary to live your life in a way you find meaningful.
High-volume, low-diversity movement/ high-calorie, low-nutrient movement:	You move a lot every day—maybe your work or lifestyle involves a lot of physical labor. You might be on your feet all day and almost never sit down for hours at a time. But the way you move is repetitive and uses the same body parts or patterns over and over again, leaving

	some of your parts strong and other parts lacking. This might describe anyone from a nurse or mail carrier to a competitive cyclist or runner who practices a lot but doesn't do much cross-training.
Low-volume, low-diversity movement/ low-calorie, nutrient-poor movement:	Most of your time is spent sitting—at home, work, and most places in between. Even if you do get up to exercise every day for an hour, the rest of the time you're back in the chair or couch. Because you don't move your whole body much, most of your parts are also not regularly involved in movement. This describes many office workers and people who drive for a living, like bus drivers.

In the following pages, I'm going to walk you through figuring out which *movement diet* you currently have and then guide you to approaching movement in a brand new way—a way that changes not only how you move, but also how you think about movement and where it "fits" into your life. As you start to become more aware of how you're moving and the choices you have in a variety of spaces, keep these movement-plan rules in mind:

Rule 1: Any movement is better than none at all.

When you feel overwhelmed at picking the "best" movement for you at the moment, remember that it's almost always better to move in any way you can. The exception to this is when your chosen activity, or your technique, creates or compounds an injury.

Rule 2: Choose movements based on your Movement Why.

It's almost impossible to determine the "best" or "right amount" of an activity, because it is almost always subjective. Certainly there are some general truths when it comes to movement; a leg exercise isn't going to train your arm, and it takes a certain way of moving your arms and legs to get your heart and lungs moving. Still, in most cases, figuring out the best exercise or program for yourself is going to require you use your Movement Why as your compass.

Rule 3: Your perfect movement plan changes with your age and stage of life.

All bodies need movement, but we are always dealing with competing needs and the pressure of too little time. A plan helps you prioritize your top needs at any given time. Children and teens, who are in the stage of building their bones enough to last them a lifetime, will have a different plan from that same person in the middle age and goldener phase of life. Injury, disease, and pregnancy are some other life stages that influence and refocus your plan.

What is your current age and stage of life?
Check all that apply.

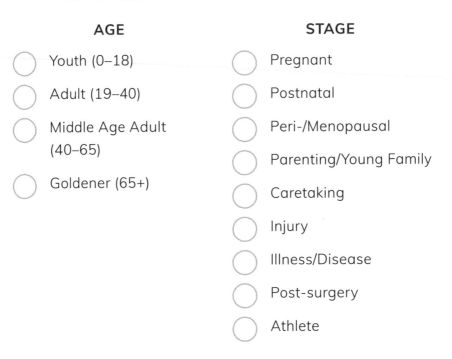

AGE	STAGE
◯ Youth (0–18)	◯ Pregnant
◯ Adult (19–40)	◯ Postnatal
◯ Middle Age Adult (40–65)	◯ Peri-/Menopausal
	◯ Parenting/Young Family
◯ Goldener (65+)	◯ Caretaking
	◯ Injury
	◯ Illness/Disease
	◯ Post-surgery
	◯ Athlete

Your age and stage will probably influence your specific Movement Why. Setting aside that we always want a body that feels and moves as well as possible and to participate in the experiences we love, make sure you're considering your age and stage as you come up with your deeper reasons for moving more.

A NOTE ON DISABILITIES

Sometimes disability is a temporary stage, as from an illness or a disease. And sometimes it's a long-term or permanent condition. Either way, disability needs to be considered as you create your movement plan. Just like different bodies have different abilities to handle different foods, there might be movements that don't work for your body. That's why we're creating individualized plans—because there is no such thing as the universal correct dosage of each movement.

If you need extra support because of a disability, fill out what you can in this book, and then seek support from your healthcare provider for more information and guidance if you're still unsure about what's possible for you. There might be some movements that simply aren't available to you, but you will also unearth movement nutrients that will be beneficial to you, too.

FIND YOUR MOVEMENT WHY

The concept of dietary nutrients arose after people realized that what they ate could affect how their body functioned. But the fact that food affects one's physiology doesn't mean that *nutrients* are the only thing people get from food. Food can also a source of joy, celebration, and tradition.

Similarly, movement is good for the body, keeps us healthy, and is key to fixing or decreasing symptoms of a wide variety of issues—but those are not the only reasons to move. Movement is the medium that connects us with the experiences and people we love. Most people want to feel better on a daily basis and are interested in living a relatively rich life, but most people also struggle with moving enough. One of the keys to moving consistently is figuring out the deeper reason behind your interest, for your general health and beyond.

Use the questions below to help you identify exactly why you are seeking out movement. As you write out the answers, keep asking yourself if you can go deeper. Answer the questions, then question your answers. Whatever you write, think to yourself, "Well, why?" to drill down to the deepest values you hold. This deepened understanding of your values and motivations makes it much easier to adapt the practice of regular movement to your life.

Think about the following activities and how your body would be used for each.

playing childhood games

competitive sports

dancing

holding children

rolling out holiday cookies

backpacking with friends

camping under the stars

taking a beloved animal for a walk

playing an instrument

planting vegetables

painting a room

exploring a new city

strolling across a sandy beach to the water

List some of your favorite, most joyful and satisfying movement experiences, and think about how your body was used. Write them out below. (I've shared a few of mine to get you started.)

Activity	Body Movement Memory
carrying my newborn out for her first walk around the block	legs walking me, arms strong holding her little body and balance! I was so mindful and thankful for feeling stable while carrying that precious cargo.
spending all day swimming in a lake in the summertime	diving and holding my breath, floating, the feeling of the water and sunlight on my skin, kicking fast, wide strokes, somersaults in the water
running 5k	pacing my breath, overcoming the urge to stop, legs feeling so strong and capable
playing "elastics" (a jump rope game) with my sister after school for hours at a time	bouncy jumps, lots of leg lifting and hip mobility, being outside
yard/garden work	lots of bending over and squatting, tool-using, sunshine, dirty hands, carrying things to and fro

Activity	Body Movement Memory

What physical experiences would you like to have that you haven't yet? What's keeping you from these experiences?

....

....

....

....

Which physical experiences do you miss, that you would like to do again? Why?

....

....

....

....

How do *you, personally*, define a healthier body?

....

....

....

....

Why does body health matter to you, personally?

....

....

....

....

Are there things you want to do with your body (a three-day backpacking trip, tying your shoes without pain, going across the monkey bars) that you can't now?

....

....

....

....

Is there a way you want to feel, mentally or physically, and you've decided moving more or differently is the answer?

....

....

....

Do you have any fears about your body's ability in your next life stage? What are they?

....

....

....

....

Do you have any fear about physical aging specifically? What is it?

....

....

....

....

How do you define "quality of life" for yourself now?

....

....

....

....

How do you define "quality of life" for yourself ten years from now? Twenty? Thirty?

....

....

....

....

PLAN POINT

Once you've spent time reflecting on the above questions and writing your answers, sum up your Movement Why in one or two sentences. This will go at the top of your movement plan.

....

....

....

....

....

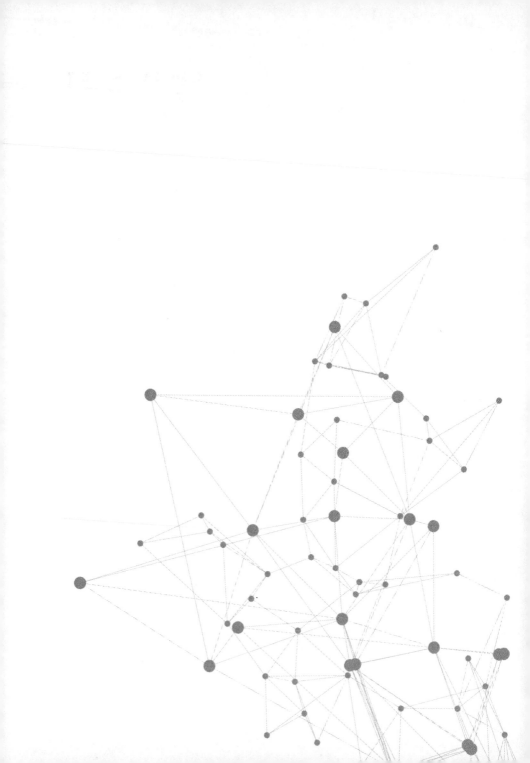

1

Movement
Calories

In this chapter, we're going to figure out how much you've been moving your body so we can determine if most of your days are "high movement calorie" or "low movement calorie."

In order to measure how much movement you're getting, we need to establish what counts as movement. As humans we can do all sorts of things with our body, and almost all of them create some change in its shape, which means almost all of the activities we do fit into the MOVEMENT category. Even getting a massage can count as body movement (more on *movement micronutrients* in Chapter 6). That said, when health professionals talk about the need for people to move their bodies more in general, they're referring to larger movements of the arms and legs, not the massage workout (which isn't a thing, but don't you wish it could be?). This larger type of movement is called physical activity.

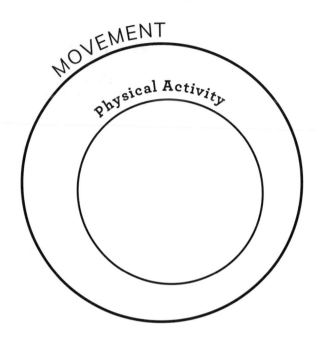

PHYSICAL ACTIVITY is a subcategory of movement and re-fers to movements that contract the skeletal muscles to a degree that burns up food energy (calories). Physical activity can be categorized by intensity: light, moderate, or vigorous. Vigorous physical activity, like going for a jog or walking up a hill, takes a lot of effort. Moderate physical activity could be something like a brisk walk, heavy house cleaning, or a game of badminton. Playing an instrument, standing at the sink to do dishes, and walking alongside a toddler are all examples of light physical activity. When we think of the types of movement that count towards good health and performance, we often think only of moderate and vigorous activity, but light physical activity is also very important to our daily movement totals. While cars driving around use the most fuel, idling cars waiting at a stop sign burn up gas too.

THERE ARE DIFFERENT WAYS TO BE PHYSICALLY ACTIVE

One subtype of physical activity is exercise. EXERCISE is physical activity that is planned, structured, and often includes repetitive, rhythmic movements. You've decided beforehand what you're going to do—cycle, walk, do yoga, do Zumba, lift weights—and how long or far and how hard you're going to do it. Exercise is also an activity done specifically to improve or maintain physical fitness.

Another subtype of physical activity is labor. LABOR is physical efforts done to produce goods or provide services. Labor might be part of someone's occupation—construction, food service, nursing, farming, etc.—but it can show up in non-professional ways too. Gardening, crawling under the house for home repairs, stacking wood, cleaning, carrying children, and hanging laundry out are also examples of labor-type physical activity.

Finally, there are ways of being physically active that don't fit into either the exercise or the labor category. Riding your bike to work (active transportation), wrestling with your kids in the living room because it's fun, going out salsa dancing, or meandering for hours through an art museum all count as physical activity.

The type of physical activity you do doesn't affect its body benefits, but recognizing all the different ways there are to be active helps when it comes to figuring out how to fit more of the movements you need into each day.

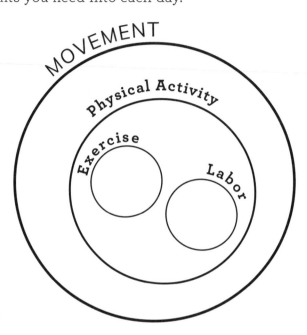

A NEW WAY OF LOOKING AT YOUR DAY

Since the Babylonians came up with the modern concept of time, we've grown more and more accustomed to thinking about each day in terms of WHEN. As you start to analyze the way you move, it is also important to think of your day as a series of WHERE, although I'm not strictly talking geographical places. Specifically, we can all sort our days into five domains: **S**leep, **L**eisure, **O**ccupation, **T**ransportation, and **H**ome—also known as the S.L.O.T.H. time-budget model.

These domains aren't exactly locations; rather, they represent where we are allotting our time. Maybe you work at home, in which case your home and occupation tasks are both at the same address. Still, the occupation tasks you do to get your work done might involve mostly writing at a computer, and the home tasks include vacuuming or stacking wood. Same address, different domain.

Don't think about how much you do them just yet; for now, list out in the spaces below the different activities you currently do throughout the week in each domain. Then note which qualify as physical activity, i.e., any movement that gets your body's muscles contracting a bit.

DOMAIN: SLEEP

While parts of our body are pushed and smushed (i.e., moved) when we sleep, this type of movement doesn't count towards physical activity.

DOMAIN: LEISURE

The amount of leisure time folks have depends on the demands of their other domains. You might have very little or you might have a lot. And P.S. Sometimes our leisure time shows up in minutes here and there rather than long bouts of time—really think about the non-essential minutes you spend looking at your phone! Here are some examples of leisure-time activities to help you brainstorm your own:

sitting/lounging   board/video games   scrolling

woodworking   social media   flower gardening

exercise (walking, running, yoga)

reading

in-person/ online shopping

music/instrument

hobbies

crafting/sewing/ knitting

puzzles

sports

watching TV/movies

My leisure-time activities

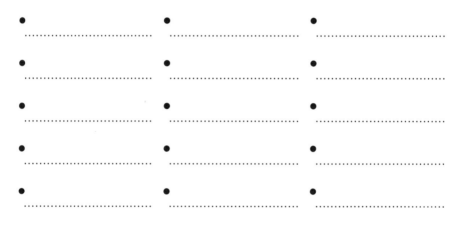

- ....
- ....
- ....
- ....
- ....

- ....
- ....
- ....
- ....
- ....

- ....
- ....
- ....
- ....
- ....

 DOMAIN: OCCUPATION

What activities fill your work time? Some jobs are many hours of the same activity and others are made up of a handful of different activities.

seated computer work

computer time

driving

kneeling and crouching

standing in place (computer, till, factory)

repetitive hand motions (keyboard, wrench, scissors, knife)

walking

lifting and carrying items

repetitive motions with heavier tools (e.g. chainsaw)

My occupation-time activities

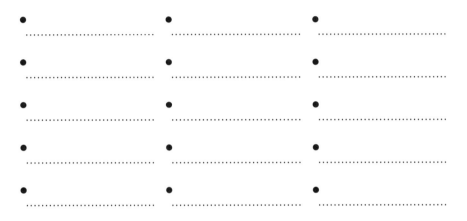

- ....
- ....
- ....
- ....
- ....

- ....
- ....
- ....
- ....
- ....

- ....
- ....
- ....
- ....
- ....

🚌 DOMAIN: TRANSPORTATION

Transportation is the action of getting to and from the places we need to go. You likely use a few different methods to commute. Below I've listed some of the different activities that go into getting around. You might have one I haven't included.

walking

train

cycling

Onewheel

elevator

take the bus

subway

rollerblade

horseback

stairs

drive/ride in car

skateboard

walking to and
from bus stop

non-motorized
scooter

motorcycle

My transportation activities

- ....
- ....
- ....
- ....
- ....
- ....
- ....
- ....
- ....
- ....
- ....
- ....
- ....
- ....
- ....

DOMAIN: HOME

All the activities for maintaining a home and the people who live
there (children, elders, animals) fit into this domain. I'm listing
a heap of activities here that are part of many households, like
vacuuming. P.S. Tools matter! Using an upright vacuum counts
as physical activity (woot woot!) but using a robot vacuum does
not. Read the tasks below to find ones that are relevant to your
life or let them help you come up with your own.

yard work (raking, lawn-mowing, shoveling)

dishes

more dishes

cleaning (vacuuming, sweeping, mopping)

washing (clothes, bodies)

stacking wood, fire building

food-growing

eating

even more dishes

animal care (dog walking, pen/barn cleaning)

caring for disabled person

paperwork/bill paying

groceries

cooking

dishes all the days

eldercare (lifting, physically supporting)

infant/childcare (lifting, physically supporting)

My home activities

- ....
- ....
- ....
- ....
- ....

- ....
- ....
- ....
- ....
- ....

- ....
- ....
- ....
- ....
- ....

HOW DO WE MEASURE HOW MUCH WE MOVE?

We live in the age of data, which means we also live in the age of data trackers. There's been a recent onslaught of wearable activity trackers—pedometers, watches, phones, bracelets, rings, and even swimming goggles—that can measure these larger movements and record how much you move. Activity trackers can't tell if you're moving because you're at work or at the gym, or which parts of the body are moving or not, but they can give you a good idea of the total physical activity you get each day.

There is also self-observation and recollection. You went for a twenty-minute walk. You rode your bike to work, which always takes half an hour. You did farm chores for two hours today. You take a Pilates class three hours a week. You walk the dog twice a day, fifteen minutes each time. You wait tables six hours a day, four days a week. Simply remembering the physically active stuff you do most days is adequate for quantifying the amount of physical activity you regularly get.

Since most movement guidelines from governing and health institutions come in "minutes per week," I also encourage folks to catalog their physical activity in terms of "movement minutes." One of the reasons we tend not to move more is because we think of exercise as being the most important kind of movement, and we believe exercise needs to come in thirty- or sixty-minute chunks to even count. Our schedule is so packed that finding thirty or sixty leisure-time minutes each week doesn't seem possible, let alone each day.

Instead, think of "movement minutes," and you'll be better able to recognize patches in your day where seven or ten or sixteen minutes are available. So in this book I encourage you to track activities by handfuls of minutes rather than in longer increments.

Revisit the activities you've listed for each domain and move the ones that count as physical activity to Chart 1 on pages 32–33. Add the average number of minutes per week you spend doing each, then answer the questions at the bottom of the chart.

THE RIGHT AMOUNT OF MOVEMENT CALORIES

We can all agree that bodies need to move, but *how much* they need to move is not entirely clear. At one end of the spectrum are very simple physical activity recommendations like "get 150 minutes of moderate to vigorous activity a week, plus some strengthening activity two days a week." Guidelines like these are solid minimums that can reduce the risk of negative health outcomes for just about every body, regardless of sex, race, ethnicity, current fitness level, or age. Still, these recommendations for physical activity are *minimums*.

Minimums are helpful given the reality of modernity, where to-do lists have become unattainably long and everything has become fast-paced. Assuming you won't be able to allot much time to moving your body, these are the minimum amounts shown by research to stave off some of the ailments that hang low on the tree of life. You might find that you're already getting the recommended weekly amount of physical activity and

CHART 1: My daily physically active activities

List physically active activities from all domains and add minutes per week.

Activity	Minutes Per Week

Activity	Minutes Per Week

WHAT IS YOUR CURRENT AMOUNT OF WEEKLY PHYSICAL ACTIVITY ON AVERAGE?

IN WHICH DOMAIN ARE YOU GETTING THE MOST PHYSICAL ACTIVITY?

IN WHICH DOMAIN (BESIDES SLEEP) ARE YOU GETTING THE LEAST AMOUNT OF PHYSICAL ACTIVITY?

yet you're not feeling great, or there are parts of your body that don't feel great.

In some cases, this is because your movement isn't distributed very well over the entire body—we'll be addressing that in the next section. It could also be that the movement minimums are not enough movement for your body.

THE "HIGH MOVEMENT CALORIE" HADZA

We can gain a deeper understanding of the impact of physical activity on the body by looking at humans who live in a way that's much closer to how all of our ancestors did—the lifestyle that shaped our heart, lung, limb, and joint anatomy into the human they are today. The Hadza are modern hunter-gatherers from northern Tanzania who have to use their bodies to gather the predominantly wild resources they live on still today, and they have no cardiovascular disease risk factors. Their diet and lifestyle are also wildly different from the diet and lifestyle of most of us reading this book.

In addition to the entire absence in their lives of processed foods and some other now-common physiological stressors like seven daily hours of screen time, Hadza spend on average 221 minutes doing light physical activity, 115 minutes doing moderate physical activity, and 20 minutes doing vigorous physical activity *every day*. That's a total of 350 minutes, or five hours of physical activity, every day. Compare Hadza "high calorie" numbers to the suggested North American guidelines of 150 minutes, or two and a half hours a week, and you'll see that there really is a huge range of being "physically active."

SOME PHYSICAL ACTIVITY IS VERY (VERY) LIGHT

Most of the physical activity minimums for health focus on moderate and vigorous movement, but to our detriment we've almost totally dismissed the contribution light physical activity can make towards the total volume of movement calories we can get each day. Said another way, our bodies need a lot of movement, but not all of it needs to involve rhythmic limb movement and burning up lots of energy at once.

The Hadza, for example, don't only get their movement calories by doing a lot of labor. Notice how much light physical activity they get compared to moderate and vigorous. Even the way they *sit* is active, too.

Think about the way you use a chair. If the back of the chair disappeared, what would happen? Would you collapse backwards? If yes, then the chair-back is doing the work of your core musculature. And obviously, if the bottom of a chair dropped out we'd fall straight down, because the chair is also doing the work of the legs.

Even though the Hadza culture is relatively much more physically active than most others on the planet, like many other cultures they also have a large volume of *non-ambulatory* (not up and moving around) time. Hadza take a lot of rest, but they and people in many other cultures around the world do what researchers call "active sitting": using one's joints and muscles to support the weight of the body while being seated.

Instead of sitting and reclining on a padded chair far from the ground, people of many cultures squat, kneel, and ground-sit.

These body positions keep muscles working at a low but constant rate for the duration of the sit, and they also require bodies to adopt different shapes (more on why that's important in the next chapter).

Active sitting is *very* light physical activity (you're just sitting there!) but still, the work necessary to hold yourself in various positions activates the cells of the body, giving them some of the nutrients they need. I mentioned before that even idling cars burn up gas. Active sitting uses energy in the same way gas-powered cars do while waiting their turn at a stop sign. Yes, it's just a little bit of gas, but over a day of driving, it adds up. Similarly, the effects of very light activity might not look like much minute by minute, but they add up. You can get more movement calories each day by turning your sitting time into *more active* sitting time. Over the last decade there's been a movement towards ball chairs or standing computer workstations in offices, at schools, and at home. These are simple ways to add more movement calories to your day without having to change the activity/task you're doing.

Next we will be figuring out the amount of non-ambulatory time that currently fills your week and how much of it is active and non-active. All of the active portion will be added to the rest of your physical activity minutes.

TAKE A LOOK AT YOUR NON-AMBULATORY TIME

The way your body feels has a lot to do with how much time you spend in place. If you're sitting quite a bit, there's a lot of potential to increase your light physical activity. Maybe you already spend a lot of time standing on your feet in place. Your activity is high, but you're not sure how to balance out that type of movement with others. No matter your personal situation, the first step is taking a good look at how non-ambulatory time shows up in your day-to-day experience.

Start by reviewing the activities you listed for each domain (pages 26–29) and move any activity where you're more or less "in place" into the boxes on the next few pages. Note if the activity is seated or standing and if it's "active" or "not active."

DOMAIN: LEISURE

During which of your leisure-time activities do you sit or stand in place? I've started with a few examples to help you fill out your own:

Activity	Sitting/Standing	Active/Not Active
knitting	chair sitting	not active
phone scrolling	couch sitting	not active
remote control airplane flying	standing	active

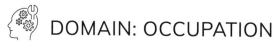 # DOMAIN: OCCUPATION

Does your occupation require lots of "in place" time? Fill out the boxes below.

Activity	Sitting/Standing	Active/Not Active
computer work	standing	active
computer work	sitting	not active
working cash register	standing	active

🚌🚲 DOMAIN: TRANSPORTATION

Transportation—to and from errands, work, etc.—often includes "in place" time.

Activity	Sitting/Standing	Active/Not Active
subway ride	standing	active
car	sitting	not active

 DOMAIN: HOME

Activity	Sitting/Standing	Active/Not Active
paperwork	sitting	not active
washing dishes	standing	active
eating meals	sitting	not active

CHART 2: Your Active Sitting and Standing Time Per Week

Copy over all your *active* sitting/standing activities and note the minutes per week.

Activity	Minutes per week

Activity	Minutes per week

IN WHICH DOMAINS ARE YOU CURRENTLY SITTING/STANDING IN PLACE THE MOST?

THE LEAST?

TOTAL WEEKLY ACTIVE NON-AMBULATORY TIME:

MAKE YOUR NON-AMBULATORY TIME MORE PHYSICALLY ACTIVE

There are many different ways to position yourself when you're not walking around. When looking at the different ways the Hadza sit, researchers Raichlen et al. found that "chair style" sitting was the least frequent way to rest and "ground sitting" the most. Squatting and assisted squatting (perching the backside on a rock or stump with the heels flat on the ground) were the most active of all the ways Hadza sit.

While you might be able to swap out some of your video-gaming, TV-watching, or crocheting positions for a supported squat, research shows there are different ways to make over more conventional ways of sitting to increase your physical activity, too. You could stand instead of sitting. Even in place, the standing parts have to work to hold you up. You could just sit in the same chair differently. Slumping your upper body against the back of the chair uses less muscle activity than scooting forward and using your core musculature to hold your own torso upright. You can make your chair wobbly. Research has shown that sitting on an air-filled cushion or swapping your chair for an exercise ball increases physical activity via more core and spinal muscle activity. While these options don't take as much muscular work as squatting, simply adjusting your position and/or what you sit upon can make your work or commute time more active.

And there are more benefits to active sitting than increasing one's physical activity: using different body shapes for an activity gives different muscles and joints a chance to move. (More on why this is important coming up a bit later.) For example, I spent very little time in "chair style" sitting while writing this book. I wrote while standing with my computer on the kitchen counter, sitting cross-legged with my computer on a coffee table, or wiggling while sitting on an exercise ball at my desk. That's one activity—computer work—but lots of different ways to do it.

ADDING UP YOUR MOVEMENT CALORIES

Movement takes up literal space, and when you start thinking about time more dimensionally, you'll find that *every minute* contains space for you to move or not. Just as a cup is a container for water, each day is a container for that day's movement. The amount of movement in a day (and the amount of water in a cup, for that matter) is called the volume.

To find your current total movement volume, add up your minutes of physical activity from Chart 1 on page 32 (activities that count as physical activity) and Chart 2 on page 42 (active sitting and standing time).

Chart 1—Physically active activities: ....

Chart 2—Active sitting and standing:.......................................

Total weekly movement calories: ....

Divide by 7 to get your average **daily** calories: ....

Place a dot where your current daily physical activity (PA) totals sit on this graph.

0 minutes 30 minutes   350 minutes
(no activity) (daily PA minimum)   (Hadza PA)

PLAN POINT

Now that you have a sense of how much of the day you're physically active, choose which one describes your movement diet:

◯ High calorie (closer to *really* active humans)

◯ Low calorie (closer to the daily minimums)

◯ Somewhere in the middle

If you need to increase your daily movement calories, the next chapter will help you figure out how more movement fits into each domain. If you're already high on daily calories, the next chapter will help you identify any sedentary domains so you can best distribute that movement over a day.

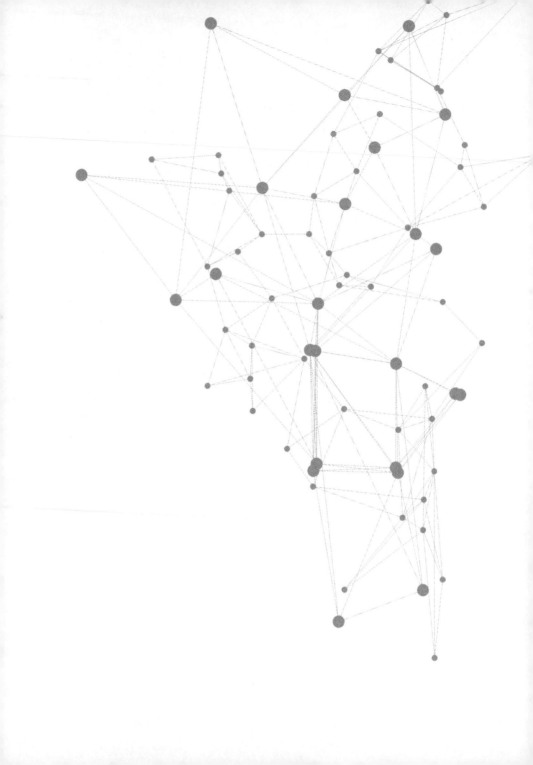

2 Filling Your Cup of Movement

"MOVING WELL" IS NOT ONLY ABOUT TOTAL DAILY MOVEMENT

Sitting down to eat one 3,000-calorie meal affects your body differently from how eating three 1,000-calorie meals throughout the day does. And both of those affect your metabolism and physiology differently from how eating ten 300-calorie meals would. You can probably guess what I'm going to say here: the same goes for movement.

How much a body needs to move is also different from *how often* a body needs to move. One of the benefits of physical activity is the movement it creates in your heart and lungs. The more vigorously you move, the more the heart and lungs change their shape, and these organs need this movement! But one of the biggest realizations in movement science recently is that regular movement isn't enough: we need to distribute movement in a way that breaks up long bouts of being still.

Of course, the first rule of movement plans applies here: **any movement is better than none at all**. But when our approach to daily movement is a single (usually leisure-time) bout, our bodies still have very long uninterrupted periods of very little movement. As a vessel that depends on movement supporting many of its functions, the human body doesn't do well being static so much of the time.

Because sedentary behavior is at an all-time high, most movement messaging has been focused on getting more total movement. There's not a lot of time for nuance—it's urgent. Still, a more complete message, rather than just "get enough movement," includes disrupting long periods of stillness. This also goes for folks getting a lot of movement when that movement is all compressed together.

What is the longest period of time—excluding sleep—that you spend in one position?

....

When awake, how many times per day or days per week do you stay in the same position for more than an hour? What activity or activities keep you in this position?

....

Many of us approach moving more solely by looking for *one* place in each day to fit thirty or fifty minutes of moderate- to high-intensity movement. As I've already noted, this is a good goal. Another good goal is finding three places for a ten-minute bout of movement. Also good is breaking up six or nine hours in a chair with getting up and stretching the backs of the legs and taking a two-minute walk around your workspace. To increase our movement calories, we need to keep all of these strategies in mind.

WHERE ARE YOU MOVING?

Unless you live a life of leisure, increasing your all-day movement requires you to look at all the domains as potential containers for a portion of your daily movement.

Based on the work you did in the last chapter, identify which scenario is most like yours:

○ I get a lot of movement calories most days, in multiple domains.

○ I get a lot of movement calories most days, almost all in a single domain, e.g. OCCUPATION (my job is labor-intense) or LEISURE (I'm a marathoner).

○ I get some daily movement calories, almost all in a single domain, e.g. LEISURE (I work out an hour every day), HOME (I walk the dog every day), or TRANSPORTATION (I ride my bike to get to work most days).

○ I don't move much in any domain.

Each of our domains has an environment, and everything in our environment affects how and how much we move. Some elements of our environment are tangible—the physical elements we interact with (a desk at school, the running shoes we exercise in, the weather). There are also intangible, invisible parts of a domain's infrastructure: the rules (your work hours), the culture (please, take a seat!), the habits (I always bike this route).

MORE THAN WILLPOWER

We come with two opposing tendencies programmed into our biology: the need for a lot of movement and the tendency to avoid it whenever possible. Human bodies evolved in natural environments where these two pulls stayed in balance naturally. You couldn't over-rest, or else you'd end up going without the stuff you need; on the other hand, moving unnecessarily wasted the food-energy that took a lot of physical work to obtain.

Over the last few hundred years, and especially in the last century, many people have begun living in new modern environments that feature numerous and unprecedented comforts and "time-saving" conveniences that require very little human movement. For all these conveniences have given us, they have also upset a millennia-old balance. When faced with the choice to move for the things we need or sit and get the things we need, we're pitted against our own very nature.

The primary reason the Hadza get so much movement is because they *have* to move to survive. The bulk of their movement is necessary labor. And their infrastructure doesn't include items like chairs and cars that reduce physical movement simply by being available practically everywhere, all the time.

The Hadza don't turn on a tap for water; they don't have one. There aren't refrigerators, stores, or DoorDash to bring them food when they want to eat. They must move—walk, run, bend, dig, carry—most days of their life. Movement, not money, is the primary commodity of nature—you "pay" for what you need with your body's movement to obtain it. The more comfortable a lifestyle we have, the more we stop moving...naturally.

Getting enough movement in modern, convenience-rich environments can require tremendous willpower, which now seems like it's gotten all tangled up with morality, but the fact of the matter is that the more you *have to* move, the more you'll move. The easiest way to significantly increase or balance your total movement minutes is to create an environment that requires you to move simply by being within it.

ASSESSING YOUR HABITAT

When creating a zoo habitat for an animal, zoologists consider all the elements the particular animal requires. The right foods, social setting, plants, water access, and movement toys all need to be available to keep the animal well. Primates need all sorts of things when it comes to movement—space to chase and retreat, three-dimensional climbing structures, toys, puzzles, and for all of these to change regularly to prevent boredom in these intelligent creatures.

Each of our domains are habitats that contain us, human animals. We too are primates.

"TIME-SAVING" CONVENIENCES

We adopt many conveniences because they save us time. However, the way many of these technologies give us time is by saving us the movement of doing a task. Drying clothes takes a lot more time when you have to hang up and take down each item individually. If you throw everything into an electric dryer, you can go do other things (maybe even movement you'd prefer to do).

As we've engaged more and more with movement-saving technologies, we've wound up in a place where the activities of daily living—the activities that once shaped and moved every body well—no longer move the body at all. And now there's no time left to move, because the day is spent operating our technologies. It turns out we didn't fill our "saved" time with our preferred movements; we stuffed it with both physical items and non-essential activities. We got more, just not more of what we need. Our need, for movement in this case, sits unmet at the end of each day because we've run out of time.

Looking closely at each of your domain "habitats," list all of the elements they contain that affect your body. Include where you are, what you wear, and what tools you use.

Note: As you make your way through this book, you will learn to recognize even finer details of your habitat that affect the movement your body is doing. Feel free to come back and add more elements to the sections below as they occur to you.

 SLEEP

The infrastructure of our sleep environment is pretty straight-forward for most of us. In the next chapter there will be more on this environment and how it moves our bodies.

mattress	pillow	bedding
futon	floor	inside
bunk bed ladder	warm	cool

My sleep environment elements:

- ....
- ....
- ....

- ....
- ....
- ....

- ....
- ....
- ....

- ....
- ....
- ....

- ....
- ....
- ....

- ....
- ....
- ....

 # LEISURE

Leisure-time activities can occur in different physical environments—reading a book at home on a couch, playing at a tennis court, heading into nature for a hike. Look back at your activities listed on page 26, then list the gear/environmental elements that they use.

bike	screen	smart phone
gardening tools	guitar	knitting needles
tennis racquet	field or court	couch/recliner
athletic footwear	dancing shoes	

My leisure-time environment elements

- ....
- ....
- ....
- ....
- ....
- ....
- ....

- ....
- ....
- ....
- ....
- ....
- ....
- ....

- ....
- ....
- ....
- ....
- ....
- ....
- ....

 OCCUPATION

On page 27, you looked at all the activities done for your occupation. Now consider and list all the tangible elements you use to do those activities. I've listed some examples to get you thinking.

desk   chair   steel-toed boots

tools   stairs   high heels

uniform   outside   inside

photocopier   water pitcher

My occupation environment elements:

- ....
- ....
- ....

- ....
- ....
- ....

- ....
- ....
- ....

- ....
- ....
- ....

- ....
- ....
- ....

- ....
- ....
- ....

- ....
- ....
- ....

 TRANSPORTATION

What are the structural elements that make up the transportation domain you described on page 28?

car seat	outside	bike pedals
inside	warm	helmet
backpack/bag	bike shoes	
bike seat	car pedals	

My transportation environment elements:

- ....
- ....
- ....
- ....
- ....

- ....
- ....
- ....
- ....
- ....

- ....
- ....
- ....
- ....
- ....

HOME

Look back at your home-based activities on page 29 and the physical elements you engage with for each of those.

kitchen sink	yard tools	cleaning tools
couch/recliner	cooking tools	

My home environment elements:

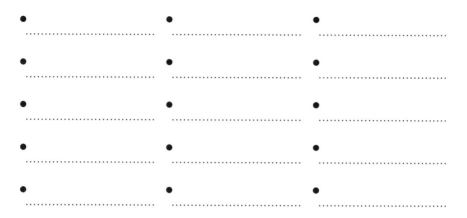

- ....
- ....
- ....

- ....
- ....
- ....

- ....
- ....
- ....

- ....
- ....
- ....

- ....
- ....
- ....

INCREASING THE "MOVEMENT CALORIES" IN EACH DOMAIN

Recall your most sedentary domains (listed on page 33). Whether you have a lot of places you could add movement or simply need to address one or two sedentary domains, there are two ways to increase movement calories: change your activities or change how you do your activities.

It's easiest to change how you do your activities. We've already discussed swapping passive sitting time for active sitting time where you can. This could be as simple as adjusting how you sit (sitting alignment) or choosing *not* to sit for an activity (e.g., stand at a workstation or stand instead of sitting on a bus commute). It could also be changing the elements of your environment: using a wobbly ball chair instead of a conventional one, or moving your computer to a low table so you have to sit on the floor or a cushion and use a different leg position.

You could also swap a typically sedentary activity for a more dynamic version of the same activity. Get out of the chair and onto the floor and sit with different leg positions while you knit. Instead of standing at the sidelines of your kid's soccer game, stretch or walk the perimeter of the field as you watch. Instead of using a vacuum robot to clean your floors, vacuum or sweep them yourself. When taking the bus or subway to work, get off at an earlier stop to fit in a little bit of outside movement before heading in.

In each of these cases, the activity—sitting at work, watching your kid play soccer, cleaning the floors, getting to work—stays the same; you've just opted for a more movement-rich version.

Changing which activities you do is harder, but possible. We all have similar categories of need that must be met regularly—to eat, move, rest, work, learn, and play, and to be with others (family and community), and in nature. Every activity we do can be sorted into these categories. Still, each of our needs could be met by a wide range of activities, so it's possible we could swap out the sedentary need-meeting activity for a different, more movement-rich activity when possible, while still meeting the same need. For example, meeting a friend for a cup of coffee and a chat can be swapped for grabbing travel mugs and heading out for coffee and a chat *on foot*. Instead of meeting your need for family time with movie nights, throw in an evening of family flashlight-tag night (and meet your need for play, too). If volunteering fills some of your leisure time, look for opportunities that add movement: help unload boxes at your food bank, donate your time to local schools as a playground monitor, or work in your community garden.

Below, look at the *non-active* activities you do in each domain and note how you can make over either an activity or the environment of the activity for more movement, or how you can replace it with a more dynamic activity. I've added examples at the top of each section.

 SLEEP

We'll set sleep aside for now. Yes, there is a way to make over your sleep-time movement calories, but I'm saving that for later. Curious? Good!

 LEISURE

The leisure domain is usually where people try to find more physical activity because, being free time, it's the most malleable. This is also the domain for play, sport, and exercise (unless you're a movement/athletic/military/law-enforcement professional, in which case those training hours are often done in your occupation domain).

Because there is the most freedom in leisure time, see if you can swap any of your non-active ways of taking leisure on page 38 with a movement-rich activity that would bring you joy.

Look back at your favorite movement memories on pages 13-14 and create a list of three or four movement-rich activities that bring you joy.

....

....

....

Review your non-active leisure-time activities on page 38. Which could be made over for more movement and/or which could be swapped for something more active (i.e., one of the movement-joys you just listed above)?

Activity	Makeover
reading in a chair	reading while sitting/stretching on the ground
reading in a chair	listening to an audiobook while walking
knitting on the couch	knitting while active sitting

Activity	Makeover
sitting and listening to music	dancing while listening to music
sitting and watching TV	watching TV while riding a stationary bike or stretching on the floor
sending memes to friends	throwing a Frisbee with friends

OCCUPATION

How can we adjust the inactive movements used in the occupation container? We can adjust our workspace (standing or dynamic-seated desks, for example) or adjust the way we organize the tasks we do in a day. Review your non-active work-related activities on page 39 and brainstorm how they might be more dynamic. I've listed a few examples below to get you started.

Activity	Makeover
sitting at a cash register	stand at register, add foot/leg stretches
meetings	stand at meetings instead of sit, take walking meetings or calls
sitting	active sitting

Activity	Makeover

TRANSPORTATION

How can you make non-active transportation time more active? Can you replace passive transportation (riding in a car or bus) with walking or cycling? Keep in mind that you can always drive partway or get off a few stops sooner, or cycle one way and get public transportation on the way back.

How much time do you spend waiting in the car? Five minutes here? Eighteen minutes there? This is a gray area—is it transportation time or leisure time? Maybe a little bit of both. Either way, I see people all the time sitting in the car and looking at their phones. Remember the minutes! This is time that could be spent adding a few steps outside or just stretching your legs or shoulders.

List the three places you go most often.

....

....

....

How could you make these over for more movement in some way? I've put some examples below.

Activity	Makeover
driving to work	walk or cycle
waiting in the car	get out and walk around, even for a few minutes
taking kids to school	drive all but the last half mile and walk from there

Activity	Makeover

 HOME

The home domain can be the fullest domain, as it includes all the tasks related to shelter, family and kids (and their schooling), relationships, and eating. The good news is that so many of these tasks can hold a portion of movement.

From ground-sitting in order to stretch your hips while paying bills, to homework walks with the kids (where you practice spelling words or memorize math facts), to stretching your calves while washing dishes (did I mention there are always dishes?)—the home domain is ripe for change!

List your five most frequent home-based tasks or chores.

....

....

....

....

....

You've just done more to increase your daily movement than most people ever do—simply by becoming aware of all the opportunities you have to move throughout the day, in all the domains. You're becoming empowered to move more, in ways that meet your needs!

How could you make these over for more movement in some way?

Activity	Makeover
laundry	hang-dry clothes, fold laundry while stretching legs on the floor
lawn-mowing	do it yourself vs. hiring gardener
lawn-mowing	push mower vs. gas-powered

Activity	Makeover

MOVING MORE AND STRATEGIC REST

Once you start tuning in to all the places movement can fit, it's hard to stop seeing them. Movement calories can be found everywhere, but does that mean we need to take every opportunity to move? What about rest?

Maybe you already get a ton of movement at work and you can't imagine tackling more movement minutes in your free time. Maybe your body is physically achy and exercise doesn't feel like the right approach. Or maybe you are mentally exhausted from spending so much of your attention—on emails, on work, on your kids—that it feels more restful to just be still.

There's a very interesting paradox going on right now: **not only are we *more sedentary* than ever before, we are also *less rested* than ever before**. How can we reconcile these two needs—for more movement and for more rest—when it seems like they're in conflict?

Many bodies need more physical rest. For example, there are people in the world who labor for work, walk for transportation, and prepare their food and tend their home without the benefit of numerous electric gadgets.

Another example of people who need more physical rest is athletes who train at such intense levels, they report feeling like they used up all their movement resources for high performance, leaving nothing until the next bout of training. This is why some of the earliest "sitting is the new smoking" research discussed marathon runners as being both highly active *and* highly sedentary, with a distribution problem when it comes to their movement—they're getting it all at once and then sitting for long stretches of time.

And then there are the many, many of us who need *mental* rest. In a high-tech world designed to captivate our attention, it's like our brain is on a treadmill set to "super fast" and almost never stops; however, our body isn't on the mental treadmill. Some of the most mentally exhausting parts of our lives don't require us to move our bodies at all. Unfortunately, when we feel mentally exhausted, we often also feel physically exhausted, and we're unable to organize ourselves into taking action. When we're mentally exhausted (and feel physically exhausted because we're mentally exhausted), it's hard to imagine that moving could help us rest; it takes so much energy! And so we continue in the same pattern of keeping the body still while continuing to run the mind on the super-fast attention-burning treadmill: texts, email, shows, podcasts, articles, reels, etc., sometimes jumping between all of these every few minutes. We think we're getting the rest we need (we're just sitting there, right?) but in fact we can be compounding the issue. Now we have an overstimulated brain and an under-moved body.

An adequate movement diet means we not only think about the entire body getting the movement (and rest) it needs, we also think about the individual parts getting what they need. If you feel like you could use more rest, take a closer look at exactly which parts of you need a reprieve.

What type of rest do you need? Is it mental, physical, or both? How do you know?

....

....

If it's physical, which parts of your body need the most rest? Which parts could still use movement?

....

....

If it's mental, which activities (or domains) require the most of your attention? Are those activities active or sedentary?

....

....

USING MOVEMENT FOR PHYSICAL REST AND RECOVERY

If fatigue of any type is a barrier for you, here's a reminder: movement is a toolbox full of different tools that can be wielded in different ways. High to low intensity; slow or fast; whole body or part by part. A wide variety of activities count as movement. Adjusting your movement diet, even adding more movement, can replenish you without depleting you.

If it's your muscles and joints that need physical rest, there are a few general approaches to using movement to make your body labor more sustainable:

Mix up your movement-types.

Break up long stretches of the same movement whenever possible. We don't have to stop moving our legs to rest our arms, and we might be able to give tight muscles on the *front* of the thighs a rest by using muscles on the *back* of the legs more. By distributing your movements differently, you can spread the movement over your body more throughout the day to avoid fatiguing any one part.

Use therapeutic exercise to maintain the areas you depend on most.

Overworked parts of our body often need healing, and one of the ways to care for stiff and overused parts is with gentle stretching or pressure (massage) movement. You can do this as your leisure-time activity (create a twenty-minute stretching routine to do daily), add these types of movements to other leisure-time activities (stretching or massaging stiff parts while watching Netflix), or you can use short two-minute breaks to tend to your body throughout the workday.

Change your form.

No matter what movement we need to do, there are options when it comes to the way we perform a physical task. By making subtle changes to the way we stand, bend, and lift, we can spread the movement in a way that allows fatigued parts to rest...even while we keep working. (Lots more on this in the micronutrient section.)

If you do a lot of intense physical labor, try mixing up activities when possible to avoid fatiguing any one particular area of the body. Package delivery is a good example of a job that mixes up movement well. Driving is punctuated with getting up and down out of the seat, some bending, some lifting, and some walking, all in the same hour. On the opposite end of the labor spectrum is bending over to plant things at ground level or reaching overhead to pick things for hours at a time. In this case, you could alternate between bending activities and reaching activities. You could also use different form to rest certain bending-over muscles: instead of bending over at the spine, you can bend over at the hips, squat, or get on your hands and knees. You can use break times to stretch muscles fatigued by bending. All of these options are different ways of moving, but they don't further tax total body energy.

Similarly, training for a sport by practicing only that sport doesn't make a sustainable athlete. In addition to cross-training, athletes are often advised to engage in "active recovery," which means moving in lower-intensity ways outside of their training time to help muscles recover. Instead of only thinking in terms of workouts, adding light, non-exercise movement to all the domains can decrease the often abundant sedentary portions of an athlete's day. Athletes might also need therapeutic movement, just as laborers do, and can always benefit from optimizing good form.

USING MOVEMENT FOR
MENTAL REST AND RECOVERY

If it's your mind that needs rest, movement can also provide what you need. You might soothe yourself with repetitive fine-motor movements like knitting—and you can do the same with big-body movements too. A meandering walk or cycle outside, especially in nature-rich spaces, does wonders for resting the mind. Many people find vigorous exercise stress-reducing too, especially when the movements are rhythmic, as in running or cycling. Practicing a familiar flow of stretches or dancing in your bedroom with a favorite song list is nourishing. Lazily swimming laps or circles, or weeding or working in a garden are all both active and restful. You can stack wood. You can carry water. These activities take very little of the type of attention demanded by our new tech-rich landscape, and they get the limbs moving while letting the mind off the treadmill.

Which movement activities make you feel mentally relaxed during or after doing them?

....

....

....

....

....

Where could these movements fit into your daily life?

....

....

....

....

Whales always need their brains working to make sure they're surfacing to breathe and staying aware of their surroundings. Instead of sleeping with their whole brains at once like we do, they shut down only half their brain at a time. One half rests while the other half monitors the needs of the rest of the body. It's so strategic! Similarly, we can be more strategic about how we allocate rest—to the parts that need it specifically.

RESTING WHILE YOU WORK

At first glance it might seem like the bulk of our resting/relaxing activities fit into the leisure-time domain, as this is the domain over which we have total freedom. However, there are ways to bring the physical activity you find restful into other domains. For example, a short walk home (all or partway) after school or work can rest eyes tired of looking at a screen, hips stiff from sitting in a chair, a body tired of being stuck inside. "Walking home" fits into the transportation domain (you're just getting from one place to an-other) but now you've increased the movement-nutrient density of that transportation activity by switching the way you accomplish getting home, and you're arriving home more refreshed for having moved. In a time when there never seem to be enough hours in a day, meeting multiple needs at once—transportation, movement, outside time, and mental rest—is an efficient approach.

As a kid I hated (hated, hated) having to do chores, but I've since learned that adding music and friends to bouts of labor makes that time not only rich in physical activity, but also feel like more like a leisure-time activity. In fact, it feels like being at a party, even if it's a movement party—more like a celebration than work.

How can we make our work feel more leisurely? People all over the world have approached their bouts of labor in this same way: taking some necessary physical task—like building a shed from a kit or shelling and pounding hundreds of nuts—and doing it in a community setting while also chatting and joking (or singing!). It's community time, movement time, and making-something-you-need time...all at once.

This passage from *Lightning Bird* by Lyall Watson beautifully captures the feeling of overlapping domains:

Each marula fruit contains a large kernel of tough woody endocarp enclosing two rich oily seeds. The seeds are small and the work necessary to extract them hard, but it is a task nevertheless undertaken, for the saying is that "mongo" is a food fit for kings. The extraction of the seeds, like the preparation of the wine, is accomplished right there beneath the tree on stones used for no other purpose. And the task usually falls to the older women, who transform it, like everything else in their lives, into a leisurely social occasion.

What are some ways you could make over your activities for more mental rest?

Activity	Makeover
cooking/dishwashing	add music and some dance steps and increase the fun and physical activity of this daily chore
eating lunch	sit outside on a blanket for some nature, hip-stretching, and eye-stretching while you eat
taking kids to the park to watch them play	jump on the monkey bars and fit your own exercises in around their play

Activity	Makeover

Now that you've taken a good look at the total amount you're moving and where you might be able to make some adjustments, the next step is to start refining your movements even more. "Calories" are the broadest way of categorizing what's in movement, and the way we check whether we're getting them is simple: is something on our body moving? Whether the movement is small or large or the intensity light or vigorous, if the position of your tissues is changing, the tissues are moving. But just like every unit of food contains nutrients beyond the calories, every unit of movement comes with more than just "we were active."

Coming up, we'll begin to explore the macronutrients and micronutrients found in movement that are always available to us, whenever we want to take them.

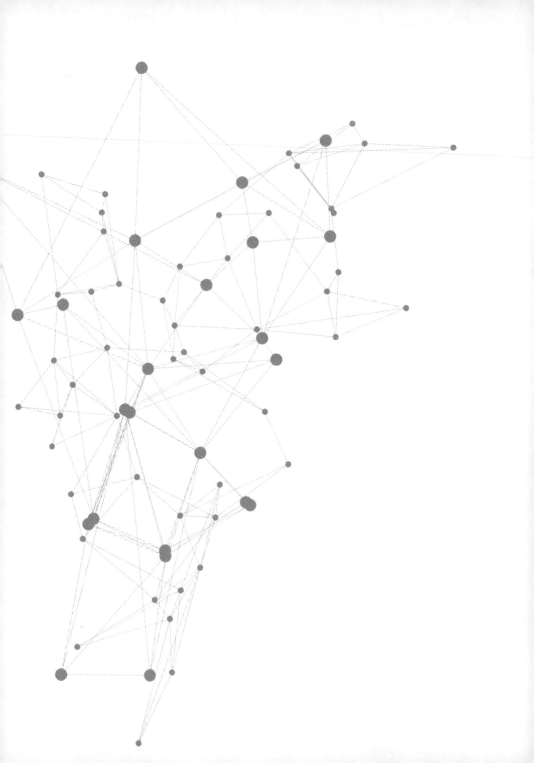

3   Your Body is a Constellation

What shape are you in right now? And no, I don't mean "looking good" or "not great" or "bad," but literally: What shape is your body making as you read this book?

In order to create your best movement plan, you have to start thinking of your body as a constellation.

A constellation is a group of stars in a recognizable pattern. Pretend for a moment that there's a star shining at each of your joints—your ankles, knees, hips, shoulders, elbows, wrists, and every place your fingers bend. Every vertebra in your spine, from the top of the neck down to your tailbone, gets a star.

Once each of your joints is shining in your mind, take a look at yourself right now. What constellation are your joint-stars creating? What is the orientation of your stars relative to each other?

How specific can you be? If you are sitting as you read this, *how* are you sitting? Is one leg stretched out straight while the other one is bent? Maybe you're perched on the end of your seat but your lower-back stars are rounded. Maybe your upper-back stars are rounded forward. Where are your shoulder stars?

Can you adjust your body so that those stars make a different constellation?

Once you start thinking of your body as a set of stars, every posture or movement now has a particular constellation or series of constellations that goes with it. The more specific you can be with imagining all of your parts as a constellation, the easier it is going to be to build your best movement program, because you'll be able to see clearly which parts of you are moving and which are not.

Just like we want to make sure we're getting enough movement throughout the day, we need to make sure we're getting enough movement throughout the body. To do this, we have to make a variety of shapes—constellations—with our body throughout the day.

WHY YOUR BODY CONSTELLATION MATTERS

All nutrients are inputs that create a chain of chemical events that change the state of one's physiology. Dietary nutrients start their interaction in the mouth, Vitamin D from the sun gets in via the skin, and all these nutrients affect cells. But *movement nutrients* get into cells via movement of the body's tissues. Moving a tissue (stretching or rubbing, for example) sets off a series of biochemical events inside the cell. There's a name for the process: *mechanotransduction* (*mechano-* meaning physical change; *-transduce* meaning to convert).

Cells matter because they are the bricks our body is built from. Every organ in every body system is made of tissues that are comprised of cells. What connects the cells to each other is a network of "extracellular matrix"—a complex network of molecules that provides structure and regulates all aspects of cell behavior.

When you move the bigger parts of your body—arms, legs, torso, head—you are rearranging the larger structures of your limbs and vertebrae, etc., and simultaneously you are also moving the small cellular structures that make up these parts. Moving the big parts moves the smaller parts inside them, too.

If you sit on a reclining stationary bike and pedal with your legs, the joints and muscles in the feet, knees, and hips get lots of movement, which in turn means the tissues in those parts are pushed and pulled. Great! You're moving, and those are minutes of physical activity. But when you consider the constellation of this exercise, you'll notice that your arm, shoulder, and core-stars aren't really moving—and neither are the cells within those parts. This type of activity is far more nutritious for the lower part of your body (where the constellation is regularly changing) than for your upper body. The cells in the upper body aren't being moved like the ones in your lower-body parts. The pedaling parts are being "fed" movement, but the other parts are not. It doesn't mean there's anything wrong with working out on a reclining stationary bike. No single food contains all the dietary nutrients we need, and similarly no single type of movement contains all the movement-nutrients we need. This is just to highlight why, when putting together your movement plan, you'll need to evaluate not only the movement calories (or amount of movement) you're getting but also how each movement affects your different parts.

Another word for *body shape* or *constellation* is *form*. If you've ever gone to a physical therapist or a personal trainer for exercise instruction, you've likely been given directions on "good form" as you worked on a particular exercise. *Squeeze your arm against your side; bring your shoulders down; shift your knees back; engage your abdominals.* These are just a few examples of what adjusting the form of an exercise might include.

Your form during an exercise is important because the angles your body makes as you move are what direct the pushes and pulls (i.e. loads) to different areas of the body. For example, doing a push-up with your hands and elbows out wide moves the muscles in the chest; doing a push-up with the hands directly below the shoulders and elbows held against the body moves the cells in the backs of the arms. Both are called push-ups, but their unique constellation makes them affect the body differently—they each provide different nutrients.

Similarly, there are subtle ways to adjust your body angles when you're picking up something heavy. One way might place more load on smaller muscles in the lower back, and the other way can place the load on the larger muscles of the legs. In both cases the activity is the same—picking something up—but by considering the constellation of each movement, you can choose the method that gives you the nutrients you need and not the ones that you don't.

If you're just moving your body to get any benefit you can (read: "calories" of physical activity), shape doesn't matter as much, but if you're wanting to increase the nutrient-density of your movement, the shapes your body makes when you move are very important. If you're in physical therapy because the way you move overworks some areas of the body while leaving other areas underworked, form is how you remedy that issue and heal. If you're lifting weights and trying to build up your body's muscle mass in a balanced way, form is also how you make sure the movement is affecting the parts you want to respond by growing stronger.

The loads movement places on the body translate into loads placed on the cells themselves, and it is at this level that body changes occur, in the form of strengths, densities, and shape. Muscle cells become larger. Bone cells stop eating up areas of bone that rarely feel weight. Skin cells increase production and build calluses on the hands to be able to keep picking up heavy tools or weight. You grow more capillaries. Your brain stores more motor programs. Yes, "the body" is getting stronger and more practiced in movement, but it's all because of *the way* each of your parts are being loaded by the movements it is doing. Coming up we'll be looking at form on a general level (macronutrients) and on a smaller scale (micronutrients). For now, I want you to get used to thinking about your body **always having a shape to it**, no matter what you're doing, and I want you to start noticing what that shape is.

THE ORIENTATION OF YOUR CONSTELLATION

Movement "gets into" your body by changing the shape of your cells, but it's not only your body shape that affects the way the cells are moved—it's also the way gravity moves the shape.

If you think of each one of your cells as a sponge, you can picture the different ways you can deform a sponge to get the water out of it. The result of the deformation is a load experienced by the cell. You can:

- » squeeze it (a compressive load)
- » pull its ends away from each other (a tensile load)
- » twist it (a torsional load)
- » slide the top of the sponge relative to the bottom (a shear load)

COMPRESSION

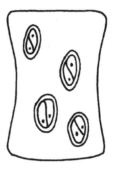

TENSION

TORSION

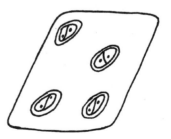

SHEAR

As you start to think about the constellation your body is making, also consider how that shape is moving your cellular "sponges." When we're trying to figure out how an activity moves us, our body shape isn't the only concern—because of gravity, it also matters how that shape is oriented to the ground.

For example, the "lying flat on your back" constellation and the "standing up straight" constellation are pretty similar. What's different is the way the cell-sponges in different parts feel the position. When you're on your back, the cell-sponges on the back of the body become compressed (that's why too much time in this position can create bed sores at high-pressure spots). But when you rotate that shape vertically, the long bones of the skeleton are now under the load of your body weight, and the cells within the bone receive signals to maintain bone density enough to support this upright weight.

This is a very real effect. NASA is very concerned with the declining bone density of astronauts spending time in zero-gravity environments. To test various exercise interventions, the agency pays research subjects to stay bedridden (here on Earth) for up to seventy days in row, in order to recreate the "zero-gravity" effects on the subjects' bone density. The *orientation of your constellation* is therefore key when you're trying to figure out which movement nutrients are created by a particular shape or movement.

SPOT THE DIFFERENCE

Consider these two images below. Both are the same constellation, but the orientation is different. How would the hands and

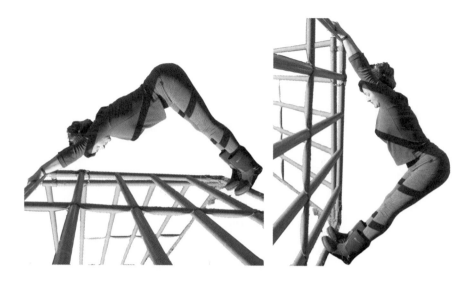

arms be loaded in each?

When the tip of the V-shape is up top (image left), gravity means the body's weight pushes—compresses—the bones of the hands and arms, and it stretches the tissues of the fingers and palms. The cells feel and bear the weight of the body.

If you rotate the shape (image right) to hang off the fence, gravity now pulls the body away from the hands. Instead of being compressed, the arms are under tension (a tensile load), from the hands all the way to the hips. Instead of bearing the downwards weight of the body, the finger and arm tissues contract against the pull of the body's weight.

Same body, same constellation—but different orientation, which means different loads and different outcomes. In addition to noticing your constellation throughout the day, start noticing the loads you feel in each of them.

Which constellation are you in most?
Describe or sketch it below.

Are there constellations you avoid due to injury or pain?
Describe or sketch below.

Which body shape would you like to be making more often?
Describe it or sketch it below.

THE HEART AND LUNGS MAKE CONSTELLATIONS TOO

Discussions about movement often involve muscle and mostly refer to the *skeletal* muscles—muscles that lengthen and shorten to move limbs and other bones around joint axes. But these aren't the only muscles in the body. The heart and lungs are made out of muscular tissue too.

Unlike the muscles of your arms or legs, heart- and lung-muscle contractions happen automatically. However, the amount of this automatic movement is fairly small. In order to take the heart and lung muscles through their full ranges of motion, you have to increase the demands on the cardiovascular system by purposely using skeletal muscles.

Consider also that arms and legs have long, sausage-like muscles, whereas the heart and lungs are each made up of a set of chambers, and every chamber is like a balloon. When you inflate a birthday balloon just a little, the rubbery walls of the balloon stretch a little. To stretch a balloon all the way out, you need to add a lot of air. The automatic contractions of the heart and lungs that happen just from being alive don't stretch the walls of these parts very much. These automatic contractions are more like putting just a little bit of air into the balloon. In order to stretch the heart and lung muscles fully, we have to do something with the entire body each day that's more vigorous than just living.

HEARTS ARE LIKE HIPS

If you've ever experienced difficulty standing up after a long car or plane ride, you're familiar with the impact that maintaining a single constellation for a long time has on your hips: they become stiff and resistant to changing their position. Maybe long trips aren't the only time your hips (or any other body part) are in the same position over and over again, all day long. Whatever the body part, keeping it in the same constellation most hours of most days of a lifetime will inevitably create stiffness! In order to continue to move easily, all parts of our body need to regularly flow through many different constellations, i.e., be taken through their full range of motion. This applies to the hips, and it applies to your heart too.

The cardiovascular system has a range of motion (see image on facing page), but many hearts and lungs spend most of their time never moving beyond their poorly inflated constellation. Over time arteries stiffen, can create high blood pressure, and begin to struggle to open when you need them to (i.e., when you want to increase the intensity of an activity). When you think of high blood pressure, recall those hips, stiff from lack of use, that can't get up out of the chair easily. Our cardiovascular system needs to be stretched! It needs to regularly get out of the meta-phorical chair and be challenged to change its shape via an increase in movement intensity.

As you work on your movement plan, you'll need to consider which of your activities currently involve different heart and lung constellations, and what you might need to add. Light activity will keep the balloon shape of these organs about the same as when you're not moving at all. Think of medium-intensity work as creating a medium-inflated balloon shape, and high-intensity work as inflating these parts all the way.

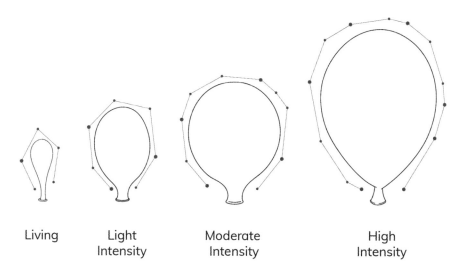

Living   Light   Moderate   High
Intensity   Intensity   Intensity

HOW TO GAUGE INTENSITY

There is no single intensity level created by any exercise, because it depends on the person doing it and their current fitness level. What might be an easy intensity for one body can be a vigorous exercise for another.

Light (sometimes called Low): steady heart rate and no noticeable change in breathing.

Moderate: your heart rate, breathing rate, and body temperature increase to the point where you might perspire.

Vigorous: feels challenging and requires exertion, heart rate increases, you may be unable to speak.

High: heart and breath rate are very high, conversation is unlikely (you're using all the oxygen you have available to move).

There are a few different ways to assess your level of physical activity. Most simply, you can do the talk test. Light intensity doesn't change your ability to talk at all. Moderate activity allows you to talk but not sing. You might be able to still talk a little during vigorous exercise, but there will often be pauses for breathing between words. High-intensity training takes almost all the effort you have, so there's nothing left for talking!

For exercise, it's often done in one-minute bouts with a rest period in between to bring the heart rate back down to a more sustainable range.

Another method based on how you feel is the Borg Rating of Perceived Exhaustion. Considering your effort, exertion, breathlessness, and muscle fatigue, you rate, on a scale from 1 to 10, how hard your body is working during a bout of movement.

Borg 1–10 scale	
Score	Level of exertion
0	No exertion at all
0.5	Very, very slight (just noticeable)
1	Very slight
2	Slight
3	Moderate
4	Somewhat vigorous
5	Vigorous
6	
7	Very vigorous
8	
9	High (almost maximal)
10	Maximal

Many devices also offer the ability to program and track intensity levels. First, you'll need to calculate your maximum heart rate: 220 − [your age]. A twenty-year-old's maximum heart rate would be 200 beats per minute, for example.

Light intensity is when your heart is beating at a rate that is 40–55 percent of your maximum heart rate. Moderate levels of activity create a heart rate that's about 55–70 percent of your maximum heart rate; vigorous activity is about 70–85 percent of your maximum heart rate. High-intensity exercise is 90 percent of your maximum heart rate.

Finally, keep in mind that our body parts (and thus whole body) get better at movement the more we practice. Today's moderate and even vigorously intense activity can become next year's low-intensity activity, especially if you're just starting to prioritize heart and lung movement. As time progresses, keep checking in with how you feel or via other tools that measure the amount your body is working.

Look back at the activities in each domain.

Which currently increases your heart/breathing rate?

....

....

....

How many minutes a week are you changing up your heart and lung shape?

....

In which domain(s) do your heart and lungs move the most? The least?

....

....

Do activities that depend on significant heart and lung movement connect to your Movement Why? If so, how?

....

....

....

Cardiovascular disease is a leader in *premature* death throughout the world, which is why Hadza lifestyles are often researched: Hadza people have healthy cardiovascular systems!

Below, place a dot where your current weekly minutes of moderate-to-vigorous activity sit on the graph below. You can include any minutes of high-intensity exercise here as well.

0 minutes
(no intense
activity)

150 minutes
(weekly intense
activity minimum)

945 minutes
(Hadza weekly
intense activity)

PLAN POINT

◯ I do almost no medium-to-vigorous activity

◯ I meet weekly medium-to-vigorous activity minimums

◯ I exceed weekly medium-to-vigorous activity minimums

100

YOUR DAILY DOSE OF MODERATE-TO-VIGOROUS ACTIVITY

While the Hadza get quite a bit of moderate (115 min) and vigorous (20 minutes) activity a day, research shows that even getting a fraction of that—20 to 30 minutes a day—can profoundly improve heart mobility and strength. You can tailor your heart and lung movement to your lifestyle and activity preference. If you like vigorous exercise you can do more, and if moderate intensity is your style, you can just do it a bit longer. You can also break down vigorous or moderate activity into smaller ten- or fifteen-minute hunks (e.g. taking the stairs when you can) or combine minutes of moderate and vigorous activity levels in one bout of movement. When it comes to the cardiovascular health benefits from getting your heart and lungs moving, duration is the most important factor—not level (moderate or vigorous).

There is no single type (mode) of movement you need to do to get your heart and lungs the movement they need. The number of activities that can move your heart and lung shape is almost endless. While it might seem that fast-paced or high-impact exercise is the only way to stretch out your heart and lungs, there are also plenty of slow, low-impact ways to create medium-to-high-intensity levels of physical activity that give the heart and lungs the movement (and thus the movement-nutrients!) they need. Shoveling show, for example, can be vigorous for the heart and lungs, even though it's slow-moving for the arms and legs. If the higher landing forces in running don't work for your knees, you can always walk uphill or carry a backpack (or baby!) to increase the intensity. Also keep in mind that everyone is starting from a different place. If you're just beginning, what might *appear* to be light activity can actually create a vigorous heart and lung response. Similarly, if you've been regularly active, you might have to start working harder to keep the heart benefits up.

Activities that get the heart and lungs moving:

» **Any "cardio" equipment at the gym**: Stairmaster, treadmill, elliptical, etc.
» **Doing any movement at a quicker-than-easy-for-you pace**: swimming, cycling, walking, jogging, dancing, vigorous house cleaning (hauling a vacuum cleaner up the stairs, shaking out rugs, etc.), lifting weights, faster-paced yoga
» **Adding incline**: walking up stairs/uphill, cycling uphill
» **Carrying weight**: walking while wearing a weighted backpack or vest or a baby/toddler/child, shoveling dirt or snow, cutting and stacking wood, tasks that involve lifting/hauling, lifting heavy weights
» **Adding jumps/ballistic movements**: jump rope, basketball, plyometric training, rebounder/trampoline (less impact)

OTHER WAYS TO MOVE THE HEART AND LUNGS

For the most part, heart and lung movement depends on movement of the rest of the body, but there are some exceptions.

Breathing exercises are a seated activity where one practices different ways to isolate the breathing muscles and lungs, and they've been around for a very long time. Even prompts as simple as "take a deep breath to calm down" can result in a conscious movement that gets the lungs moving beyond their small, at-rest levels of inflation.

Getting into a sauna is another way to get the blood vessels to open and your heart pumping without needing movement of the rest of the body. Heat is an environment that creates a physical response, and there's more and more research on the heart-healthy benefits to this very old practice. If you're wondering if sauna-ing is right for your body, check with your doctor about any contraindications.

What's interesting about both breath-exercise and sauna movements is they're both a form of relaxation that gets you moving. If you identified you could benefit from some rest in Chapter 2, these are both "dynamic relaxation" options, with breathing exercises slotting easily into multiple domains.

Which heart-and-lung activities appeal to you?

....

....

....

....

....

Which domains can you add minutes of moderate-to-vigorous activity to?

....

....

....

....

....

CREATE YOUR OWN GALAXY OF MOVEMENT

When it comes to movement nutrition, the way you're doing your activities really matters. Your whole physical experience alters when you change how you hold and move your body,

DAMP CHAMPS AND SWEATY BETTYS

While heart-and-lung-moving activities can often fit easily into different domains (leisure, home, transportation), one tricky thing about this type of activity is that it can be harder to slot into our day seamlessly—especially when it leaves us in need of a shower or change of clothes afterward.

When it comes time to build your movement plan, keep this in mind: what can often work is getting your heart-and-lung constellation changes in first thing in the morning, near the end of the day, or when going pre- or un-showered is sufficient. Save the easier moves for when you need to stay fresh as a daisy.

because that affects how the joints, muscles, cells, and all the outside forces relate to each other. Our bodies are nourished by making a whole galaxy of constellations!

We've already explored how you're doing your activities in terms of intensity (light, moderate, vigorous, and high); next we'll be exploring the impact your shape—your constellation—has on different parts of your body, and how to thoughtfully balance a movement diet.

4 The Movement Macronutrients

For thousands of years, humans have been trying to figure out what's in food and how what we ingest affects the body. With time, and microscopes, we've learned more about the elements that make up food. First we understood the larger *macro*nutrients—fat, protein, carbohydrates—then, one after another, came the *micro*nutrients. Diseases once believed to be caused by a virus (pellagra), a toxin (beriberi), or the environment (scurvy) turned out to be symptoms of nutritional deficiencies. It was in trying to solve these public health issues that the concept of a nutrient emerged. Only through a long period without certain inputs—such as sailors experience at sea without fresh food—do we come to recognize them as essential.

The macronutrients within food are impossible to see with the naked eye, so how do we make sure we're getting a good variety of dietary nutrients? With food groups! Food groups are collections of foods—like "meats," "dairy," "fruits and vegetables"—with similar macronutrients and appearances. Food groups are a slightly expanded version of the protein, fat, and carbohydrate categories, created to make eating nutritiously easier for everyone. By getting roughly the right amount of each food group, we can build a well-balanced plate.

Just as each food has its associated body-benefits, every way we use our body—the constellation it makes and amount it works—nourishes us in a specific way. Again, this is why getting "enough physical activity" isn't a complete movement plan. To get a well-balanced movement diet, we need to watch the shape of the movements we do with our body each day to make sure we're regularly fulfilling all the bigger categories of movement.

You've probably already heard of cross-training. Cross-training developed because many athletes were logging multiple movement-hours in their sport and only their sport. While practice made them better for a while, at some point performance began to decline. It wasn't necessarily that they were getting too much practice, but that by only training some of their body parts over and over again, they were inadvertently creating areas of relative weakness.

This is an inconvenient truth about nutrients: there's an appropriate dose for each, and they need to be in relationship with the right amount of other nutrients. Spinach is a nutrient-dense food, but it doesn't contain all the nutrients we need. If your diet is mostly spinach, you will become malnourished over time, overloaded in some nutrients and deficient in others. You need other foods too. Movement works in the same way. Doing the same movement over and over again, even if it's a good-for-you-and-you-love-it kind of movement, can leave some parts overworked and some parts underworked. We have to regularly move in different ways to fit in all the different movements our body needs.

WHAT ARE THE MOVEMENT MACRONUTRIENTS?

Food guidelines help folks get the right balance of the macronutrients—protein, carbohydrates, and fats. Conventional physical activity guidelines for health take a similar approach by suggesting weekly movement be broken down into three primary categories of movement: *cardio* to move the heart and lungs, *strength-challenging* activities for muscle, and *stretching* for joint mobility.

There is wisdom in these recommendations, as they make sure you're distributing movement over *tissue types*. This recommendation also indicates that experts already know movement calories aren't the only way we should think about a movement plan: *different tissues* need different types of movement to be well. But, because the benefits to movement are local (they occur where the body parts are actually moving), we need to take things a step further: a complete plan needs to make sure movement is also distributed over all our body parts. We need to also consider *tissue location*...which brings us back to considering shape.

THE SHAPE OF HADZA ACTIVITIES

Let's consider the movements the Hadza perform.

All transportation is walking, usually while carrying something ranging from very light (a spear or empty basket) to heavy (a child, a load of wood, two buckets of water). Hunting, also mostly walking, could involve a short sprint and archery—the pull of a bow and the making of all hunting tools.

Digging up food from the ground requires crouching; moving on the hands and knees; and the repeated grasping, arm movement, and shoulder movement used to work a hoe or stick into the earth—similar to the movements found in tending a garden, or pulling up a potato harvest, only with more squatting.

Gathering fruit and honey from trees requires hand and shoulder stretching and strength for climbing. Cutting wood, building and tending fires, and cooking food all require a multitude of movements for each step. Building shelters takes more walking, more bending, more squatting, more handling of tools.

Sometimes there is a little dancing, too.

Then there is all the active sitting. Just like many of our days, a significant portion of a Hadza day (and those of other hunter-gatherer groups) is spent *not* moving bodies from point A to point B. The difference is that the resting positions used by cultures around the world come in many more shapes than "chair."

We could loosely sort these hunter-gatherer activities into intensities or into cardio, strength, and flexibility categories, but if we really want to understand the way these movements affect the cells of the different parts of the body, we have to sort them into categories that consider shape: upright locomotion, carrying, climbing/clambering, fine-motor making, big-body work, and active rest.

Movement Macro 1: Upright Locomotion

Hadza walk a lot and run a little. Recent step-count research shows Hadza take around 15,000 steps a day on average. While the mileage depends on the individual, it's safe to say Hadza walk about six miles a day, more or less. Although walking and running are unique activities that use the body differently (more on movement micronutrients coming up), they both have somewhat similar shapes and loads. The legs carry the weight of the upright body while they (and sometimes the arms) repeatedly alternate back and forth.

Other activities in this group have a similar general constellation: rollerblading, ice skating, cross-country skiing, and hiking. Exercise machines like the elliptical and Stairmaster are also designed to create gait-like movement nutrients.

Movement Macro 2: Carrying

Carrying motions add weight to the body. Whether you're carrying in your hands, in your arms, on your head, or attached to the front or back of your body, it increases the work of the heart and lungs plus whatever muscles you're using. There are many different carrying shapes—more on that below—but in each case the body is upright while under an extra load. While the Hadza might spend some time holding a load while in place, carrying is often done at the same time as upright locomoting.

Other activities in this group include backpacking, waiting tables, rucking, stocking shelves, moving boxes, carrying or wearing kids, "farmer carries," etc.

Movement Macro 3: Climbing/Clambering

This is a big group of motions in which your hands, arms, and shoulders carry a lot of your body weight. The individual constellations can vary. Maybe you're hanging from your arms; maybe you're crawling up a steep hill on your hands and knees. In all cases, your arms are loaded with some, a lot, or all of your body weight and you're often doing some sort of moving forward that depends on them.

Other activities in this group include crawling under the house, paddling out to surf, swimming, exercises like pull-ups and chin-ups, rock climbing, roofing, climbing trees, and anything that gets you on your hands and knees.

Movement Macro 4: Making Movements

Hadza, and many cultures the world over, spend lots of time making the things they use. This is a group of lower-effort movements that keep the hands, arms, and shoulders moving in smaller but somewhat rhythmic ways while the rest of the body is still, as when you're making tools, instruments, fires. Hadza often pair making movements with active rest positions, and you can too.

Other activities in this group include woodworking, basket weaving, processing food by hand (think shelling peas), sewing, beading, playing an instrument like piano or violin, auto or bicycle repair, petting or playing tug with an animal, ceramics, tending plants, styling or cutting hair, putting away groceries.

Movement Macro 5: Big-Body Work

This is another big group of motions that use the body in a moderate to vigorous way while the body is more or less in place. Just as the "protein" category of food can be filled with beans or meat, there are numerous activities other than walking, climbing, or carrying that require the bigger muscles of the body to work rhythmically. Consider hoeing, digging, and repeated picking up of a load (think shoveling).

Some activities in this group are raking leaves, shoveling snow, building, massage/bodywork (giving, not receiving), moving house, shaking out rugs, lifting weights, and repeated bending and lifting.

Movement Macro 6: Active Rest Positioning

This group of movements includes a wide variety of body constellations, but with the exception of standing rest positions, they involve being on the ground or floor in sitting or squatting shapes that take the leg joints through diverse angles and ranges of motion.

Other activities you could put into the "active rest" group could include floor work portions of yoga or mobility exercise

classes, playing games on the ground with kids, using a squat toilet, and using a standing desk or floor workstation.

The Hadza, and any hunter-gatherer culture, are not living relics. Still, their activities of daily living are much closer to the ancestral activities that shaped our anatomy. While our activities of daily living are perhaps quite different from those of the Hadza, our anatomy still requires us to use our bodies in these shapes. We can skip them and still survive, but, like if we're consistently avoiding an essential dietary nutrient, malnourishment will show up at some point in our tissues. Thus, we can use these general body shapes as our movement food groups—our movement macronutrients or movement macros, as I'll call them going forward—as we look to see what our movement diet contains.

The good thing about a food group approach is it leaves us free to pick the foods from it that work best for us. If you hate brussels sprouts, you can pick a different brassica you prefer. If you don't want to eat cheese, you can also get calcium from almonds. Similarly, there are many activities that fit into each group of shapes, and we can build our movement plan out of the ones that suit us best.

WHICH MOVEMENT MACROS DO YOU HAVE, AND WHICH DO YOU NEED?

Consider your list of weekly activities in each domain below and list those that fulfill any of the movement categories above. I've included samples for each domain to get you thinking.

 LEISURE

Activity	Macronutrient(s)
Backpacking	upright locomotion and carrying
Kettlebell session	big-body work
Playing guitar	making movement

OCCUPATION

Activity	Macronutrient(s)
Floor desk	active rest
Hair cutting	making movements
Delivering mail	upright locomotion + carrying

 TRANSPORTATION

Activity	Macronutrient(s)
Walk to work with backpack	upright locomotion + carrying
Load bike on bus	big-body work

 HOME

Activity	Macronutrient(s)
Garden planting	big-body work
Yard raking	big-body work
Meal-making	making movement
Wrestling with kids	big-body work

My movement diet contains the following movement macronutrients:

- ◯ Upright Locomotion
- ◯ Carrying
- ◯ Climbing/Clambering
- ◯ Making Movements
- ◯ Big-Body Work
- ◯ Active Rest Positioning

As we've come to outsource more and more of the movement
that goes into living, you might find that your current activities
of daily living aren't moving your body very diversely. There
might be entire movement groups missing. Not to worry. Missing
nutrients are commonplace with food diets and with movement
diets too. Learning what we are currently getting and not getting
is an essential step to building our movement plan.

PUTTING IT ALL TOGETHER

You'll recall that every dietary nutrient has a "right amount."
When it comes to movement, we likely have more freedom than
we do with dietary nutrients, but to get a sense of how the volume of the Hadza food groups relate to each other, consider this
food-pyramid-inspired graphic of a Hadza movement diet.

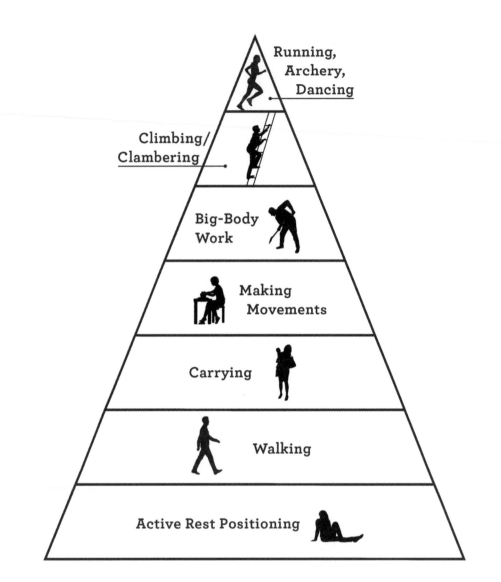

Movement Macros Pyramid

When we think of hunter-gatherers, it might conjure up images of high-speed running or throwing spears. While these movements are likely essential to the lifestyle, they are the "peak" movements ("top of the pyramid") because they are relatively infrequent and not the movements that fill large portions of the day for most of the population. Hadza get a ton of movement, but most of it might even look boring. There's a lot of quiet, stationary activity and also lots of pretty basic walking. Good news, this also means a lot of it is accessible to the lifestyles many of us live today—no sprinting or spears required. *Can* your plan include sprinting and throwing spears? ABSOLUTELY! It's your Movement Why and it's your plan. But just remember this: the bulk of the movement calories making up a Hadza movement diet probably look pretty humdrum to many.

Food pyramids are a general approach to eating a more balanced diet, and movement pyramids are the same. Also, pyramids will change depending on your age and stage of life, so keep all that in mind. Right now, we're just learning how to think in terms of the way we're using our body.

If you were to sort your activities into the movement macronutrient categories—walking, running, carrying, big-body work, making movements, and active rest—and organize them by amount of time spent in each, what would your current pyramid of weekly movement look like?

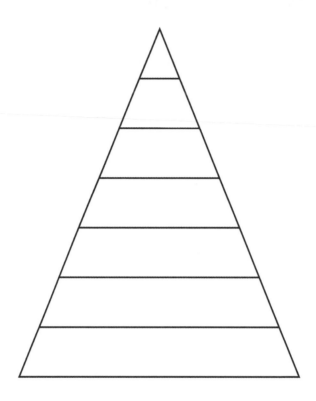

Is your pyramid looking a little unbalanced? Don't worry. Below are some ways you can start adding missing macronutrients to your life.

MORE ACTIVE REST POSITIONING

In previous chapters, we've looked at the physical activity created by active sitting and squatting—specifically the muscular work used to support your body in these positions—but this isn't the only benefit. Modern chairs prevent our ankle, knee, and hip joints from flowing through a variety of different shapes. So much of what we do to "stretch" our legs can occur simply by mixing up the way we sit, adjusting our constellations. Also, getting down to and up from the floor is a weight lowering and lifting exercise; your entire body is the dumbbell (no offense).

How often do you sit on the floor or squat?

....

Does moving down to and up from or being comfortable on the ground relate to your Movement Why? If so, how?

....

....

....

If you need to increase your volume of active resting positions, here are some challenges that can help you change your habits. They also make this practice fun for the entire family (i.e., they're not about Mom or Dad doing exercise, but just mixing up the regular routine. Kind of like sneaking vegetables into kids' food!).

Ad-Break Boogie: This can be fun and boisterous. Everyone can get up for a run, or a quick game of up-and-down-off-the-floor, to see who can do it the most before the show you're watching returns.

Couch-Free Week: Challenge the family to get OFF THE COUCH for a full week, choosing the floor (and some pillows if necessary) instead.

Indoor Picnic Night: Spread out a blanket or tablecloth and eat on the floor. This one's so easy, I recommend doing it at least a couple of nights a week. It feels like a holiday!

Chore Squat: For a month, every time you're folding laundry, cleaning the floor, gardening, or doing other work that could be done seated or on all fours, squat for as long as you can to do the work. You may need to stop squatting periodically, but once your joints are ready, get back into a squatting position.

Floor-Based Workstation: Here you can sit or squat as you work at a low table. Take your newspaper, laptop, and reading time to this area.

Make a Squat Toilet: You can use a product like the Squatty Potty, a homemade platform, or a couple of overturned coffee cans. Getting yourself into this configuration regularly for elimination will benefit your intestine and will also get your joints more used to the squat position.

MORE WALKING

Walking is kind of like a superfood because it gets just about all your parts involved when you're doing it, plus it's a movement that's very easy to pair with non-movement tasks. It's great as transportation, for getting together, for adventuring. It's also free.

What amount of walking (length or time) counts as "taking a walk" to you?

....

How often do you walk on average, and how far?

....

In which domains do you walk? Do you usually walk for the same purpose? (This could be purely for exercise—leisure; to get groceries—home; as part of your commute—transportation, etc.)

....

....

....

How does upright locomotion relate to your Movement Why?

....

....

....

....

If you need to increase the amount you're walking every day, here are some challenges to try.

Get a Walking Buddy: ALL the recommendations below will be easier to facilitate if you go get a walking buddy, right now.

Start Every Morning with a Walk: This walk can be whatever distance you can manage—a couple of times around your home, once around the block, five miles.

Walk After Every Dinner: Again, any distance you can manage.

Up Your Mileage: If you're not walking every day yet, start walking every day, and work on increasing your mileage gradually but steadily until you're walking at least three miles every day, whether it's all at once or in smaller chunks. If you're already there, work up to five or six miles daily. Make walking part of your life: walk for groceries, walk for coffee, walk to meet friends, walk with your friends.

Never the Same Way Twice: Mix it up! Every day for a week, change your walking route. Even if it's a walk with a purpose, like dropping your kids off at school, running errands, or walking to work, find ways to vary the route. The change can be as small as walking on the other side of the street, but it's even better for your brain and body to explore alternatives that involve other paths and roads altogether. This will likely have the bonus of increasing your walk time a bit, since chances are you currently choose the most direct route!

Pick Your Kids Up from School on Foot: You don't have to walk the entire way, but park a bit away, giving you and the kids a chance to log some extra steps.

Mega-Walks: Choose one day a week to walk double your average single-walk distance; choose one day a month to attempt to walk triple or quadruple the distance. Bring snacks, water, audiobooks, and company as needed. (Bonus tip: Locate public bathrooms or private bushes beforehand.)

MORE CLIMBING/CLAMBERING

Like all movements, upper-body movements in which you support your weight with your arms come in a variety of shapes. Not all of these movements will be available to you now, and some may never be. You can still make sure your movement diet has at least some movements in this category.

How regularly do you get onto your hands and knees?

....

How often do you carry a load (groceries, child, books)?

....

Could you carry very heavy loads such as furniture for a short distance?

....

How often do you hang from your hands?

....

Can you go across the monkey bars? If so, how do you move across the bars? (One at a time? Every other?) How many bars can you move across before you have to stop? Which body part requires you to stop first?

....

....

....

....

....

How does hanging or carrying strength relate to your Movement Why?

....

....

....

....

Here are some challenges to increase this macronutrient.

Doorway Walkthrough: Start by regularly getting your arms overhead. Every time you walk through a doorway in your home, reach up as high as you can (taller people will likely be able to reach directly up; shorter folks, reach as high as you can to each side of the doorframe, one arm at a time). Hang out here for a few breaths.

Playground Progression: Create a walk (you're taking daily walks, right?) that takes you to a local playground's monkey bars or by a tree with branches you can reach. Start hanging with two hands, then practice one-handed hangs. Once you can do these with ease, work up to swinging back and forth and, if you're on monkey bars, try reaching for the next bar.

Ditch the Grocery Cart: The next time you pop into the grocery store for a few things, don't get a cart or a basket. Carry each item you need in your arms to the checkout. If you're there for a big shop with your family or a friend, spread the work around—kids, spouses, and friends can all benefit from using their bodies too. Bonus: You're likely to find that big shopping sprees aren't conducive to this task, so there'll be a natural shift to smaller, more frequent trips to the store, facilitating regular walking.

For Those with Kids or Pampered Pups—Ditch the Stroller: Not all the time. Not for a ten-mile hike, at least not yet. But if you have a child who can walk but usually rides in the stroller, start taking walks with them. Be willing and ready to carry them for short bouts—on your back, your shoulders, front, hip, etc.—while they develop the strength they've been out-sourcing to the stroller. It's great not just for them to begin walking more, but also for you to be able to bear the weight of your little/medium/big ones. Remember, when the going gets tough, the tough bring bribes (i.e. snacks).

Hanging Bar in the House: This is probably the smallest change you can make, but it's so effective. Install a chin-up bar over your most high-traffic doorway, and instead of hang-ing out in that doorway à la Doorway Walkthrough, actually *hang* there every time you go through. You can start working up to chin-ups/pullups here, or swing, or just hang from both arms, or one arm. Again, the point is to do this every time you go through the doorway—think of it as a magic password.

Monkey Bars in the House: If you've already gotten rid of your couch, then I know almost for certain that you have room for a five- or six-foot set of monkey bars where the couch used to be. If you don't have that kind of space, consider hanging bars from the ceiling along a frequented hallway, and never go down that hallway without traversing the bars (once you're able to). Use your living-room monkey bars to hang while you're watching TV or listening to podcasts or visiting friends.

Get Gleaning: Of course you can always just go climb a tree, but if you're looking for additional purpose, join a local fruit gleaning group and put those arms to work for your local food bank!

MORE MAKING MOVEMENTS

There are tons of activities that can get you your making movements, and they're easily tailored to something you like. If you are also looking for a way to fit in more rest, this type of movement can often also be restful for the mind while it keeps the body in a light state of activity and exercises the making parts of your body and brain.

How often do you move your fingers on the screen of your phone?

....

Do any making-type activities appeal to you now? Have any ever appealed to you? What are they?

....

....

....

How do making movements relate to your Movement Why?

....

....

....

Here are some challenges that can help you increase your making movements.

Motorless Mondays: Once a week for a full day, avoid all motorized kitchen tools. Grind your coffee beans, chop and mash your smoothies, handwash your dishes, slice your veggies yourself instead of using a food processor.

Traditional Cooking: Find an old recipe from your family line (or one you find interesting) and spend a day making it from scratch, paying attention to how much movement goes into the process. Even our recent ancestors moved a lot more than many of us today.

Darn It: Pick up a new fine-motor hobby you can combine with some active sitting when you're resting: darning, embroidery, crochet, etc.

Whip It: Next time you whip cream or meringue, get the fork or a whisk and do it by hand.

Write it Down: Write a long note to a friend and see how quickly your hand starts cramping. We've even lost those super-fine motor skills! See if you can write a few paragraphs with the other hand too.

MORE BIG-BODY WORK

There's a perception that we lose strength as we age, but really we just stop doing the tasks that require muscle. One fine way to get this movement macronutrient is to lift weights at the gym. Another way is to look around for activities that challenge the body to work—they're more plentiful than you might think.

Are there any big-body work movements you used to do but stopped doing? Why did you stop?

....

....

....

....

Are there any big-body work movements you're currently outsourcing that you could reclaim?

....

....

....

....

How do big-body work movements relate to your Movement Why?

....

....

....

....

Chore Restore: Reframe chores as exercise sessions. I hated chores as a kid, but once I realized that in addition to getting a necessary task done I'd also be getting some of the movement I needed, I enjoyed them a lot more. Set a timer, turn on your favorite music, and get that big-body work done more happily.

Garden Workout: Join a community garden, not for the plants or people but for the workout. There is big-body work available in an assortment of constellations, and the good news is that you don't need to be responsible for more than accomplishing your physical work—the garden and the garden manager are like a free personal trainer.

Be the Neighborhood Helper: You know these people—the group everyone calls when they need help moving an appliance or loading a truck. You can also formalize this kind of helping labor by volunteering in a food bank warehouse or helping with the big cleaning jobs at animal rescues, for example.

Say No to the Blow: If you can't imagine a task without your power-blower, at minimum do a human-powered portion of every job—sweeping, raking, or shoveling—first.

Use Your Kids (or Small Friends) as Weights: Not all movement has to travel somewhere. Swoop kids up and spin them around daily. Let yourself be a human jungle gym and connect with your strength and younger kids at the same time. Practice some acro or partner yoga with your preteens and teens as a way to get some fun contact alongside your movement. Make muscles and memories.

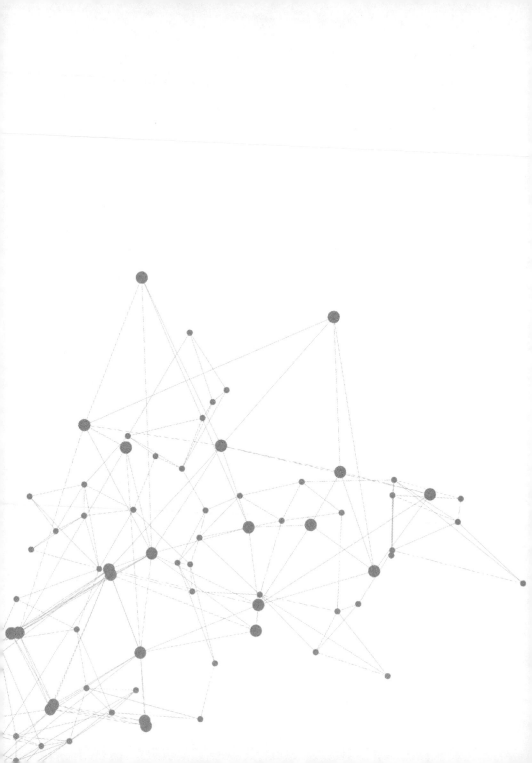

5

Purposeful, Meaningful, and Functional Movement

While we need to sort movements into macronutrient shape-categories to understand and adjust our movement diet broadly, a successful movement plan requires other ways of sorting activities too. Other important distinctions we can make when it comes to movement are purposeful movement, meaningful movement, and functional movement.

Purposeful movement meets other needs beyond increasing individual physical fitness or health. These activities can increase physical fitness and health, but that's not their only purpose. Walking the dog because it needs exercise, carrying a baby because it needs touch and transportation, digging a garden to grow food, riding a bike to the store as transportation, stacking boxes at work, and splitting wood for heat are all examples of purposeful movements. In contrast, exercise programs are often made up of *non-purposeful* movements: walking, stepping, or cycling on machines that take you nowhere; lifting weights for no practical reason beyond the effects of lifting them; doing bending, twisting, and reaching stretching exercises without any immediate task requiring you to bend, twist, or reach.

Categorizing a movement as purposeful or non-purposeful says nothing about how it affects our health and wellness (both types work!) or whether it should be included in our movement plan. In fact, non-purposeful movements are often necessary to balance our movement diet in a modern context. However, non-purposeful movement can only fit into the leisure domain. If time is a hurdle for getting the amount and shapes of movement you need, purposeful movement—movement that can be done while you get something else done—is the ticket to increasing your total movement minutes.

Meaningful movement holds personal significance for each of us and is perhaps even tied up with our individual or group identities. These movements are part of "who we are." You decide which activities fall into this category for you, and anything goes. Surfing, backpacking, playing soccer or pickleball, long-distance walking or running, sitting in meditation, working in the garden, riding horses, dancing, or hours practicing hand movements on the piano or guitar all count. There is no distinction here between purposeful and non-purposeful movement. One of your meaningful movements might be running on a treadmill because it makes you "feel like yourself." Your meaningful movement might even trump typical health pursuits like physical longevity. Athletic pursuits like high-level sport, performance art (e.g. dance, circus), or potentially hazardous occupations (e.g. firefighting, military), all of which can cause a great deal of wear and tear on the body, could relate to your Movement Why.

Some might argue that *meaning* isn't important when it comes to movement nutrition—it should be all about cells, muscles, and health variables. But to stick with something, we need it to be meaningful to us. This is true for building a nutritious food diet—making sure we enjoy and connect to the foods we put into our bodies, personally and culturally—and it's true for movement too. Only about a quarter of Americans meet the current CDC guidelines for physical activity, a fact that plagues exercise adherence researchers. Why aren't people exercising when they know all the benefits to physical and mental health? When they know it makes them feel better? I believe a primary reason is that non-purposeful movement—which, again, is what constitutes most exercise programs—simply isn't meaningful to many people.

How did we get to a place where exercise is the primary way movement is prescribed? Perhaps that happened because exercise is likely meaningful to those who study and then research movement, create movement guidelines, and give expert guidance on physical activity. The meaningfulness of exercise *to them* is what drew them to study the topic in the first place. P.S. I put myself in this group! Exercise is where I personally found meaning, which is what took me to university to study it for both undergraduate and graduate degrees, and continues to keep me excited enough to write books about it. But I can also see that just because *I* find exercise meaningful doesn't mean that exercise is the best approach for everyone else. Exercise is only one type of movement, and it's mostly a non-purposeful, leisure-time pursuit. There are other categories from which we can build our movement diet. Every movement doesn't have to be meaningful, but if we want more bodies getting the movement they require, meaning is essential to building a movement plan. If you've got ample leisure time available, your movement plan might include lots of exercise, but just know that there are other ways to build out the movement you need.

Functional movement is a particular approach to exercise that prioritizes movements that help the body stay strong enough for the activities of daily living. If you were to consider what your body might need to do on a daily basis throughout a lifetime, it would include combinations of movements like getting up and down out of a chair, reaching the arms overhead, carrying small loads, bending over to get something from the floor, twisting your spine to look over your shoulder when backing a car up (I know, I know, cars now have a camera and an alarm that tells you if you should look, but maybe also keep turning to look just in case), balancing on one leg, changing direction quickly.

There's a principle in exercise science, the Law of Specificity, which describes that the way the body responds to movement is very specific to the activity itself. For example, sitting and pedaling a recumbent bicycle won't improve your standing or walking balance. If you want to improve your standing or walking balance, you have to practice activities that include you challenging your balance while standing or walking. If all your yoga poses have you *pushing* your arms into the ground, your yoga practice will not improve your pulling strength. If you want to improve your pulling strength, you need to practice movements that include elements of pulling.

Functional training has you evaluate how you use your body when being physically active (so back to those shapes again) to make sure that you're already regularly practicing the movements you want to be good at later.

A functional training approach is a good way to focus your movement plan, and it also relates to *purposeful* and *meaningful* movements. Many of your activities of daily living are purposeful movements that you'll ideally keep performing over your lifetime. And hopefully the movements that bring you the most meaning are also part of your regular activities. Perhaps they're not daily, but you'd want to also ensure that your movement diet helps sustain these "most important to you" ways of moving too.

List some of your purposeful physical activities

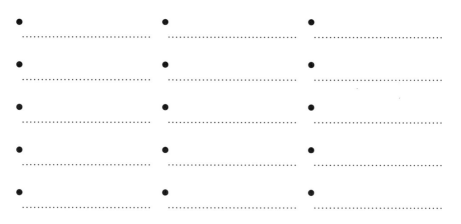

List some of your non-purposeful physical activities

- ....
- ....
- ....

- ....
- ....
- ....

- ....
- ....
- ....

- ....
- ....
- ....

- ....
- ....
- ....

PLAN POINT

Let's hone in on your most meaningful movements and their place in your movement diet and your life.

Which activities are meaningful to you?

....

....

Is there any overlap between your *purposeful* and meaningful activities?

....

....

Any overlap between your *non-purposeful* and meaningful activities?

....

....

....

Do your purposeful and meaningful activities have anything in common physically, emotionally, spiritually, culturally? If yes, what?

....

....

Do your meaningful activities relate to your Movement Why?

....

....

What constellation/body parts are used in your most meaningful movements?

....

....

ACTIVITIES OF DAILY LIVING, PHYSICAL FITNESS, AND FUNCTIONAL TRAINING

I'm not sure how the Hadza would answer questions about meaning and purpose when it comes to their movement. The bulk of Hadza movement is purposeful because it has to be. When you need to move for your essentials, there's probably little time and energy for non-purposeful movement. And hunter-gathering movements are likely meaningful to hunter-gatherers because they are culturally relevant. They're part of the identity of a hunter-gatherer. Further, Hadza movements are all functional because every day they practice the movements they will depend on daily throughout a lifetime. It's all very efficient and a contrast to the scenario many are dealing with now, where very few activities of daily living are active, purposeful, and meaningful.

LABOR IS NOT THE OPPOSITE OF LEISURE

Labor falls in the category of purposeful movement—you're producing goods or services. While it's often associated with generating income, a lack of options, and even oppression, labor doesn't have to relate to any of these concepts. Labor can be enjoyable, for the physicality of it as well as the mental rewards (I'm looking at you, dopamine!) that come from completing a purposeful physical task. Whether we outsource it or not, labor is a required element of most domains, and many people have labor-rich hobbies—crafting, building, gardening—that fill their leisure time. Many people seek out occupations that require more movement—waiter, package deliverer, landscaper, nurse—so they can simultaneously meet their need to earn a living plus their need for movement (we all know, or are, people who "just can't imagine sitting at a desk all day"!). It's even possible to think of movement as one form of "payment," especially if it's well-balanced.

So labor need not be seen as negative. Adopting these purposeful activities could allow us to get the movement equivalents of calories and balanced nutrients that we require.

Where could you add or reclaim labor movements in the non-occupation domains? I've started with a few examples for each.

Sleep Labor

- Rolling out and picking up floor bed daily
....
- Setting up a tent
....
-
....
-
....
-
....
-
....
-
....
-
....

Leisure Labor

- Throwing pottery
....

- Woodworking
....

- Sewing or knitting clothes
....

-
....

-
....

-
....

-
....

-
....

-
....

Transportation Labor

- Create a "walking school bus" for your neighborhood
....

- Car/bicycle maintenance (change oil/chains, tires/inner tubes, etc.)
....

-
....

-
....

- ....
- ....
- ....
- ....
- ....

Home Labor

- Carrying groceries home
  ....
- Baking bread
  ....
- Gardening
  ....
- ....
- ....
- ....
- ....
- ....
- ....

THE PHYSICAL ACTIVITY PARADOX

In general, the benefits of any physical activity are positive, no matter how you get it done. However, research has shown there is something called the Physical Activity Paradox: high physical activity demands *in the workplace* can negatively impact physical and mental health. This seems counterintuitive. Why doesn't worktime activity affect the body in the same positive way leisure-time activity does? And it's maybe even more counterintuitive when you consider the Hadza and the fact that the bulk of their movement would be considered labor, which is the opposite of leisure, right? (It's not, but hear me out.)

The research that made these findings compares labor (occupational physical activity) to exercise sessions (leisure-time physical activity). Exercise sessions tend to be shorter in duration (thirty to sixty minutes), be higher in intensity, include unrestricted movement and body positions, and incorporate sufficient rest between sessions. In contrast, occupational movement tends to require longer sessions of low-to-moderate activity made up of pre-determined and repetitive movements. Leisure-time and worktime physical activity also differ psychologically: leisure comes with more choice and control. Notably, Hadza tribes have no boss, their repetitive activities are often turned into social gatherings, and their movement repertoire is diverse.

All of this points to the fact that getting "enough" movement cannot be the entire movement plan. For years, researchers have tried to figure out combinations of movement to make occupational movement less detrimental, and the results vary, often depending on the situation. Sometimes increasing movement in the other domains helps—*if it's meaningful*. Adding a walk to work along a route made stressful by cars and pollution might not work; however, a food-making party with family, cleaning/organizing your home in a way that pleases you, or tending a garden can reduce some of the negative impact of occupational labor. It's also been shown that some occupational tasks begin to exceed workers' physical capacity as they get older and their fitness level declines. While the Law of Specificity should keep your heart strong enough to haul boxes at work if you haul boxes at work, it's not the only principle at play. As we get older, we have to work harder to keep the same fitness levels. Research has shown that many folks with active occupations perceive they're getting enough exercise at work and don't supplement in any other domains. In this case, a functional training approach would include adding "non-purposeful" training exercises to strengthen the body to prepare it better for the movements needed at work.

Which domain is most labor-rich for you?

....

Describe your occupational movement. How many minutes each day of high, moderate, and low-intensity? What shapes of your body do you use?

....

....

....

What other movements might balance your occupational labor?

....

....

....

WHAT DOES IT MEAN TO BE PHYSICALLY FIT?

The field of exercise science defines physical fitness as "the ability to carry out daily tasks with vigor and alertness, without undue fatigue and with ample energy to enjoy leisure-time pursuits and to meet unforeseen emergencies." Said another way, you need the strength, endurance, and mobility to get your daily work done without it beating you down to the point where you can't enjoy your off-work time. And you need to keep your body resilient enough to deal with unexpected events that might demand more physicality than your daily activities do—dealing with injury, illness, or natural disaster (flood, earthquake, fire, pandemic).

As more and more people develop lifestyles that require very few *active* tasks of daily living, we are becoming less fit. We might miss this at first because we're able to continue to meet these lower-intensity daily tasks—think using computers and phones for hours every day—with relative ease. And we still have energy for leisure-time pursuits...right? Maybe? See the previous section on mental fatigue and why sitting around to do computer work isn't leaving us with the physical energy to do much else with our bodies, despite our daily work being poor in physical activity. And this is all before the inevitable unexpected event arises.

If we gauge fitness only by our ability to do modern daily work tasks instead of what humans have done for millennia, it's easy to get stuck in a fitness blizzard and not see where we started and where we're going. That's why considering movement macro-nutrients helps: they are an anchor of sorts to where our bodies came from. They reflect the tasks human anatomy has been built to perform. We can use this framework to select tasks or activities that are available to us now, in our current environments, making sure we also include those that bring us meaning.

Then we shake all of this down through the movement micro-nutrients to see which holes are left in our movement diet.

6

The Small but Mighty Movement Micros

Picture the constellation Orion, standing with one arm raised and wielding a sword, the other arm stretched forward with his shield, and three stars making up his belt. Now zoom in to look at just the belt. This collection of three stars is an *asterism*—a smaller observable pattern/group of stars that aren't a whole constellation. If whole-body constellations make up different movement *macros*, then smaller observable groups within those constellations make up "movement micronutrients," which are like the minerals and vitamins found in movement.

You can eat adequate calories and have a good balance of carbohydrate, protein, and fat but still be missing a necessary vitamin or mineral and end up with a health problem related to malnourishment. Health issues can also be created by getting *too much* of a particular micronutrient. AND I BET YOU

KNOW WHAT I'M ABOUT TO WRITE NEXT: The same goes for a movement diet. We need to pay attention to our micros and observe all the smaller portions of the body to see how all areas or parts of the body are moving.

There are usually a few spots in our body not moving at all, and these areas of *sedentary cells* are often why our body hurts or doesn't function well in certain areas or ways. In food diets, micronutrient deficiencies can occur when we don't consume enough food in total, if we don't consume enough of the foods that contain all the vitamins and minerals we need, or when we can't absorb the nutrients from the food. Similarly, we can be movement-micronutrient deficient when we don't move enough in total, if our total movement isn't diverse enough in shape, or if "sticky spots" in the body prevent our movement from getting "into" all the areas of the body.

STICKY SPOTS

Sticky spots are areas in the body where, instead of a body part gliding easily at its joints, it becomes clumped together with parts around it.

Say we are looking to see if your movement diet includes Vitamin Get Your Arms Overhead (a necessary motion to keep the shoulder moving well). At a quick glance, this movement vitamin is easy to spot: look through your daily activities and mark down the ones that get your arm(s) moving overhead.

But we can look closer for more information. Our bodies are good at accomplishing a task however they can. Imagine your shoulder joint is really stiff—maybe you use it repetitively at work, maybe you had surgery in this area, or perhaps the bulk of your activities keep your arms down by your side. For whatever reason, one of your shoulders resists moving in certain directions. But you need to get your arm overhead—to serve a tennis ball or write on a chalkboard or paint your bathroom. What many bodies do in this case is arch the lower back to lift the chest, which allows the arm to raise *without* needing the shoulder to move very much. The overhead-arm task has been completed by "going around" the stiff shoulder. The arm and the chest stay clumped together into a sticky spot and the shoulder is under-nourished (it's not moving much), while the lower back becomes over-nourished with *too much* Vitamin Lower Back Arch.

Fast forward in time. The joint of the stiff shoulder might hurt from disuse, even when you're not moving it, or perhaps you've stopped moving it altogether because that's how it feels best. The lower back aches because it's been compensating for your shoulder. But your pain doesn't make sense to you, because this bypassing of shoulder movement happens subconsciously. After all, you can *see* your arm going overhead so you think you've got the micros all balanced out (it's not related to immobility because, look, my arm goes overhead no problem!) and the lower back has also become extra mobile, so there's also no movement problem here that you can easily spot. If you decide to keep studying more advanced nuances of form (see page 252), you'll learn to watch how the shoulder itself moves, rather than looking just at whether the hand lifts.

Patches of sedentary tissue can develop via lack of use, habitual movement patterns that leave out a body area, or injury or disease that temporarily or permanently prevent an area from moving. Whatever the cause, these under-moved areas often interfere with our overall movement because they end up hurting or keeping us from moving our whole body more.

In this chapter you'll be breaking down your whole-body movements into smaller pieces, but don't confuse the *smallness* of what you're looking at with the size of its importance. Movement micronutrients are powerful and are my favorite to teach; seeing the nuance in how you move is a giant key that unlocks movement mysteries you might unknowingly have.

Even if you already get a lot of daily movement because your occupation, sport, or toddlers keep you moving, **movement micronutrients can be key in figuring out why a body part or two (or three) doesn't feel good or move the way you want it to.**

If you *don't* get a lot of movement because an area of your body hurts or isn't moving well, **movement micronutrients can be the solution to getting your whole body moving more**.

Finding the movement micronutrients you're getting too little or too much of can be transformational to how your body works and feels.

WHAT ARE THE MOVEMENT MICRONUTRIENTS?

Every cell of the body needs movement, but we can't put a star on every cell—there are trillions of them. How can we narrow down which constellations—or asterisms—we need? I've found people can make tremendous improvement in how their bodies move and feel by looking closely to see if their movement diet contains the elements listed below. Some are joint positions, some describe a way of moving the body that creates a particular set of necessary tissue forces, and a couple are environmental scenarios that directly impact the shapes we get into.

Seated/Hip Flexion	Heart Movement
Hip Extension	Lung Movement
Dynamic Hips	Dynamic Eyes
Dynamic Knees	Pressure Movements
Active Feet	Upright Weight-Bearing
Fixed Spine	Upright Balance
Dynamic Spine	Agility
Grip Movement	Impact
Forward Arms	Community
Dynamic Arms	Outdoors/Nature
Core Movement	

This list is not exhaustive, but it addresses the micronutrients that create some of the most common issues in the body over time when our daily movement contains too little or too much of them.

In the next section, you'll be evaluating your own activities and which micros are "in" the movements you do most regularly. Read the following descriptions and refer to them for clarification when needed.

Seated/Hip Flexion

When you sit, your hip joints move to bring the thighs up, closer to the torso. This action is called hip flexion. Many people get far too much of this movement micronutrient, which in overly abundant quantities is associated with muscle tensions that affect how the knees, hips, and lower back move the rest of the time. Examples of activities that would check this box: chair sitting, driving, cycling, rowing, horseback riding, kayaking.

Hip Extension

Hip extension is the opposite to hip flexion: the thigh moves back behind the torso. Hip extension uses the gluteus maximus muscles—your butt muscles!—which then go on to support the lower back, pelvic floor, hips, and knees. Examples of activities with hip extension: walking, running, cross-country skiing, yoga warrior poses, hip-flexor stretches, swimming (crawl, backstroke).

Dynamic Hips

Hip joints do more than go straight forward (flex) and backward (extend): your legs can move out to the side and across the body, make circles, and rotate. You could eventually break this micronutrient up into each of its directions, making sure you hit them all, but for now, mark this box if your activity moves your hips in any direction that isn't forward or backward. Examples: swimming (breaststroke), dancing (hula, salsa, etc.), "hip opening" yoga poses or stretches, certain meditating positions, active sitting positions, squat variations, martial arts, soccer.

Dynamic Knees

Knees do more than straighten and bend; they rotate a bit too, and they are built to carry our body weight up and down regularly. Mark this box for activities including or similar to: active sitting positions that bend your knees fully or require different angles than sitting in a chair does; hiking or walking on complex terrain; negotiating hills, stairs, and ladders; getting down to and up from the floor; treading water; squat variations; being on your hands and knees; certain meditating positions; doing things at ground level.

Fixed Spine

Some activities call for your spine to stay more or less in the same position, and often this is a forward-rounded shape (think hunching the upper back forward). How often does this micronutrient show up in your daily activities? Examples of fixed-spine activities: reading, being on the phone, cycling, dishwashing, driving, computer work.

Dynamic Spine

Our spine is made up of vertebrae, parts that each move in multiple directions: nodding forward and backward, tipping to the right and left, and rotating. This is what allows our spine as a whole to bend forward, arch backward, bend sideways in both directions, and twist! When we don't use all the ranges of motion throughout the spine, we eventually stiffen and lose the ability to. Activities that use the spine dynamically include golf, Frisbee, house cleaning, yoga, Pilates, putting dishes or groceries away, specific spine stretches.

Grip Movement

How often do you grasp—a tool, a tree branch, a rung of a ladder, a barbell? Activities that would check this box include: woodworking; gardening/farming; rock climbing; home/car repairs; vigorous cooking (moving heavy pots and pans, chopping, kneading, etc.); mountain biking; bowling; pullups; using hand-held weights, stress balls, or a grip-strength exerciser. Make an extra note if your activities involve gripping only with one hand; you'll want to work on making sure both hands are getting some Vitamin Grip!

Forward Arms

Many activities fix our arms out in front of us. Because our eyes are in front of our head, a lot of our arm work will be out in front of us. We get a lot, and I mean a *lot,* of Vitamin Forward Arms, and too much of this shape can create muscle tensions that go on to affect how the arms and shoulders feel and work. Examples of activities that would check this box: computer work, device holding, driving, reading, bicycling.

HOW TO FIX A FIXED SPINE

There's nothing inherently wrong with fixed-spine activity if you're not spending too much time in that position. That said, the shapes our spines tend to be fixed in have changed recently. Just take a quick glance around any public place and check out the number of heads looking down. Not only looking down with the eyes, but with the head. Not only a little nod of the head, but a big forward curve in the neck and upper back. Not only a few people, but many—maybe even most—people.

Looking down for certain activities is nothing new. Reading, sewing, cooking, art—these are age-old activities humans have had to look down for. It's an important movement and part of why our spine is made of separate hinges instead of one long, straight bone. These spinal hinges (our vertebrae) curve into the looking-down shape quite easily, as if we were made for this movement. What is new and does pose a problem is the amount of Vitamin Dropped Forward Head we're getting and how little we do any other movement of the head, upper back, and even upper body. The micros are not well-balanced in this area.

You can change the loads (and thus the movement micros) without having to move any other part of your body, just your head, with this Head Ramp exercise supplement. You can use it as an exercise or try it out as a new position when you're getting too much Vitamin Dropped Forward Head (you know you can lift your phone up rather than dropping your head down, right?).

Start by touching the top of your head.

Without lifting your chin or your chest, reach the top of your head up toward the ceiling while sliding your ears back over your shoulders, as shown in the photo on the right.

The head ramp movement is the opposite of the forward head shown on the left. It undoes a forward head by picking your head up away from the ground, moving it back over your shoulders and lengthening your neck.

This adjustment doesn't negate the need for dynamic spine time too; it just increases the nutrient density of your fixed spine time.

[Adapted from my book *Rethink Your Position*]

Dynamic Arms

Shoulder joints allow our arms to create almost an entire globe of arm movement. Arms go out in front (you already knew that) but also side to side and overhead and behind us—and combinations of all of these, too. They can also push and pull in each of these positions. Our shoulders need these movements to stay fully nourished. Some activities that check the "dynamic arm" box include swimming, hanging laundry, certain yoga poses, gymnastics, painting (walls, not pictures), rock climbing, playing violin (cello, etc.), paddling.

Active Feet

There are thirty-three joints in each foot, and these all need to be moved in order to stay moveable. Stiff shoes can prevent movement of many of these parts, as can not being on your feet often. Sometimes certain areas of the feet are hypermobile and move too much, which can also create a problem in the feet. Activities that move the parts of the foot include barefoot time, barefoot walking in sand, walking or running in minimal footwear (toe-spreading space, flat and flexible sole), climbing (rocks or trees), certain yoga poses, martial arts, foot exercises, foot massage.

Core Movement

Our core muscles move whenever our spines are dynamic (see Dynamic Spine), but additional ways to engage this area of the body include using the arms for carrying and lifting, paddling (e.g., kayak and standup paddleboard), hula hooping, sitting on a ball and other active sitting positions, and core exercises.

Heart Movement

In this category, list activities that increase the beats per minute of the heart. Remember to include short-duration activities like taking a flight of stairs, or running across the street to catch a bus. Heart-moving activities can be created by increasing the intensity of anything you're already doing.

Lung Movement

Lung-moving activities increase the depth or rate of your breath. This can be done by increasing the intensity of movements you're already doing or via breathwork exercises.

Dynamic Eyes

Eyes have muscles too! Just like hips can get stiff sitting in the same position hour after hour, eyes get stiff from looking at the same distance too long. They have ring-shaped muscles that tighten when you look at things up close and widen when you look at far away things. Looking at devices, reading books, knitting, and woodworking keep those muscles tight (as does sleeping); looking up from a book to see across a room opens them a bit; looking far away out a window, across a yard, down the street, to distant mountains, etc., opens them a lot. Our eye-focusing muscles need to be taken through this range of motion, but these days, more people spend a lot of time contracting their eye muscles to do "near work." Mark this box for activities that require your eyes to focus beyond near-work (and even better, beyond the wall of a room) like: driving, baseball, tennis, golf, archery—most sports, really—bird watching, stargazing, hiking, taking a walk outside, looking up from this book and taking an eye-exercise break at a window!

Pressure-Deforming Movements

It's easy to see the squats or leaps in a hunter-gatherer's movement diet. It's harder to spot smaller movements that don't look as exercise-y. If we think back to the Hadza and the movements they get, a frequent movement not yet mentioned is the way the surfaces of various body parts are squashed against different natural surfaces while just living life. Many of our environments have been made artificially flat or cushioned, effectively removing this subtle but abundant movement. Activities that create pressure-deforming movements include receiving a massage, foam rolling, fascial-tissue ball work, playing on the floor with kids, climbing trees, bar work (gymnastics, fitness), sitting on the ground, lying/sleeping on the ground.

Upright Weight-Bearing

Human bodies are heavy, and when standing upright we place our full weight on our skeleton. This upright posture is essential to keeping our bones and joints regularly loaded—especially the hips. Mark this category for any activities that have

MORE ON PRESSURE-DEFORMING,
FLESH-SQUASHING MOVEMENTS

I've struggled with what to call these types of movements—movements of the skin and fascia and adipose and muscle that don't involve contraction, just the squashing and tugging of flesh as a body part is pressing against something else. All I've been able to come up with for now is "pressure-deforming movement" which is pretty meh. But more important than what we call it is identifying when we see or feel it. Arm tissue smushing into limbs and bars as you hold on to or climb them; the pre-callus (or pre-blister!) tug of a bar or shoe on your skin; leg-tissue deforming to accommodate walking over, sitting on, or lying down on the bumpy ground; and the push of sleeping surfaces on our bodies are all examples of micromovements that affect the state of the body.

Most of the surfaces our bodies interface with have been made artificially smooth or cushioned. We sit and lie on cushions most of the time; the shoes we wear have become thicker and thicker over time, so we can't feel the texture we walk over. At the same time many are turning to pressure-deformation movement supplements—massage, body/ball rolling, cobblestone mats—to replace the movement nutrients that come from moving your body atop/over/through objects that move a lot of smaller parts when you interact with them.

By turning our habitat into flat, smooth, and cushioned surfaces that are so unlike Hadza terrain, we've inadvertently eliminated hundreds of little but important movements. We've reduced our body contact points to our shod feet (mostly), hands (a little), and our knees (hardly ever). If you couldn't see how exercise practices like foam or ball rolling related to hunter-gatherer movement, maybe you'll be able to now. These are movement vitamins to replace some of the movements that unmodified or lightly modified landscapes have always required of humans as they move through them.

you stacking your body weight vertically. Standing "Big Body Labor" macronutrient activities (e.g. shoveling, standing and lifting weights), hairdressing, walking, running, rucking, racquet sports, grocery checking, standing at a work desk, washing dishes, and bowling all count. Are you standing? Mark this box!

Upright Balance

The ability to stabilize yourself when upright is essential to maintaining your ability to be in upright weight-bearing positions, which is essential for bone health. Activities that challenge your ability to balance while you're upright are key to helping you stay upright comfortably throughout a lifetime. Walking, running, surfing, dancing, skateboarding, roller/inline skating, slacklining, paddle boarding, or using balance beams and balance-training boards can all mark this box.

Agility

Movements that mark the "agility" box require the whole body to make a rapid change in direction while traveling. Rapidly changing directions loads joints abruptly and can often lead to injuries if you don't regularly get enough Vitamin Agility. Practicing these abrupt joint loads trains the smaller parts of our body to better resist damage from movements that occur unexpectedly. Many sports require agility, including racquet sports (pickleball, tennis), as do wrestling with a dog, trail running, playing tag, waiting tables or other occupational activity with sudden pivots, and agility-training exercises.

Impact

Impact is the jolt of force created when your body moves away from the ground then hits the ground, usually with your feet. Bones and joints respond well to regular jolting, but there's a point where too many jolty loads break down tissue instead of building it. The same goes for the pelvic floor—another place that receives the impact of the landing (the organs hit the pelvic floor when your feet hit the ground). Impact that's irregular or extremely forceful can be injury-making, but it's health-making to do regular, low-volume impact practice that builds up body parts to become stronger. Movements that create impact: running, landing jumps, hopping down off something, jumping rope, tennis, gymnastics, burpees, jumping jacks, some styles of dance. Lower-load impact forms, especially helpful during periods of injury rehabilitation, are running laps in a pool or running/jumping on a trampoline.

Community

When it comes to movement, social support is very helpful. The benefits of community—not just having other people around you, but doing something with others you feel you *belong* with—is health-making. A social group can keep kids and seniors more active overall. For parents of little ones, having community is essential when it comes to modeling movement for kids plus moving more throughout the day. In middle age, having friends who value movement makes it easier to get moving yourself. (Maybe you're the friend who is making movement more inviting for others!) Adults spend twice as much time in nature when they have friends or family who make a concerted effort to get outside. Check in to see which activities deepen your sense of community or belonging. Fill in this box for any activities done with others, like walking with a buddy, sport practice or theater rehearsal with your team or group, cooking in the kitchen with loved ones, or going out dancing with your friends.

Outdoors/Nature

Being outside moves us differently and more broadly than being inside does. The texture and terrain of what's underfoot, the varying temperature your body deals with, and natural light and ample distance to look into all create unique movements. Mark this box for activities that get you out of the house! If you're on a patio, in a yard or park, on a mountain, or walking down the street for your activity, mark this box!

TRACKING YOUR MICRONUTRIENTS

The following pages contain micronutrient trackers, one for every domain. Each row represents four hours of a week. On each chart, you will **list activities you do regularly** in the left-hand column. Consider how many hours you spend each week engaged in that task—and the constellations you use for that task. When you're done, you will be able to see any full or missing columns and have a sense of the more specific ways your activities are and aren't moving your body.

As you list activities for each domain, keep the following in mind: Be specific. When you're thinking of your activities, break them down into tasks of different shapes or body use. For example, if nursing is your occupation, instead of listing eight (or twelve) hours of "nursing" as an activity, identify your most occurring tasks that make up your nursing time: seated deskwork, *standing* computer work, lifting patients, hallway walking, etc. This will give you a clearer view of all the different ways your body is being used at work.

In some cases, activities in one domain that are very similar in both purpose and shape—like watching a movie, watching TV, or phone scrolling—can be listed under a single activity name, like "sitting and watching entertainment/media (TV, movie, scrolling)."

If your activities vary significantly by season, your micronutrients likely do as well! You can fill out separate sheets for different times of year. If you have more activities than can fit onto a page, or you are revisiting this book and need new trackers, download sheets at nutritiousmovement.com /MPMPresources.

A NOTE ON SLEEP SHAPE

We sleep almost a third of every day, thus the time spent in our sleep shapes really adds up. Regardless of what we're doing for physical activity, we are also moved by how we sleep! The tracker for this domain will be easy to fill out; there isn't a wide variety of micro-movements you can get in your sleep. Sleep is when you stop moving and your body recovers from what you did all day. That being said, many of us sleep in one or two shapes, and on the same surface, over and over again, and our bodies adapt, just as they do when we're awake. Many of us sleep in a certain shape—hips flexed, spine fixed and rounded forward, arms out in front—that looks similar to the computer work position used for many occupations, and the looking-at-media shape that makes up a lot of leisure time. Again, there aren't many options for different movement nutrients during sleep time, but seeing how sleep time stacks up with non-sleep time is important when it comes to understanding your movement diet and where it might need adjusting.

MICRONUTRIENT TRACKERS

» **List your regular activities in the ACTIVITY column for each domain.**

» **Every row is worth 4 hours. Fill in the micronutrient boxes for that activity, one box for every 4 hours. For activities you do fewer than 4 hours a week, use one row and fill in the smaller quarter-boxes for the appropriate micronutrients: one quarter-box for 1 hour, two quarter-boxes for 2 hours, three quarter-boxes for 3 hours. (See Sample 1.)**

» **Activities you do for longer amounts of time, e.g. 8, 20, or 46 hours a week, will use multiple rows. (See Sample 2.)**

» **Once activities are listed and micronutrient boxes filled, add up the number of (4-hour) boxes for each micronutrient, and write in the TOTALS line at the bottom, e.g. 8, 4.5, 2.25, etc.**

» **Follow the instructions at the bottom of each tracker to easily transfer your Most, Moderate, and Minimal micros to the chart on page 190–191.**

Sample 1 has been filled out with a few tasks for each domain so you can see some examples of which micronutrients I've filled in.

SAMPLE 1: WEEKLY MICRONUTRIENT TRACKER

ACTIVITY	Seated/Hip Flexion	Hip Extension	Dynamic Hips	Dynamic Knees	Active Feet	Fixed Spine	Dynamic Spine	Grip
SLEEP								
At home on mattress	■					■		
Camping (sleep on ground)	■					■		
LEISURE								
Hiking (hills)		■	■	■	■		■	
Pickleball game			■	■	■			■
Watching media: movie/TV/computer/phone scrolling	■					■		
OCCUPATION								
Serving food		■	■	□	■	■	■	■
Grocery stocking			■	■		■	■	■
Deskwork (seated)	■					■		
Deskwork (standing workstation)				■	■			
TRANSPORTATION								
Drive to work	■					■		
Walk partway to work		■	■		■		■	
Bicycle to market	■	■	■		■		■	
HOME								
Meal prep			■	■		■	■	
Cleaning (sweeping)			■	■		■	■	■
Walking the dog		■	■	■	■		■	

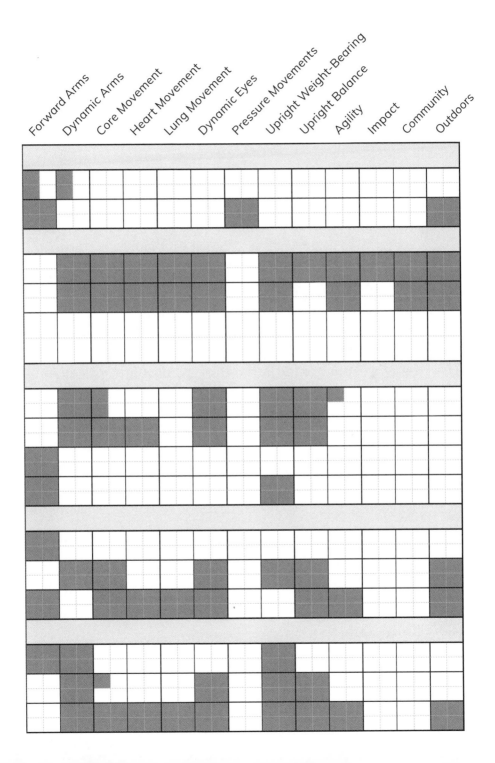

For domains like occupation and sleep, where we do the same activity over and over again for many hours, it's possible you will have the same activity listed all the way down the left-hand column. In some cases, your activity might not move you exactly the same way. Maybe your spine is fixed practically one hundred percent of the time while you sleep, but your arms are overhead part of the time; maybe some of your sleep is on your side with your hips flexed and some is on your back with your legs straight. Your work might be almost entirely at a desk, but you do some of it standing and some sitting on a ball. The micronutrient boxes you fill in will vary based on the shapes you're using.

The example opposite is of a Tracker for the Sleep domain. The sleeper's estimates for an average 7-hour night's sleep are 2 hours side sleeping, 2 hours sleeping on the back with knees bent, 2 hours sleeping on the stomach, and 1 hour sleeping on the back with arms overhead. Each nightly 2-hour position translates to 14 hours per week, or three full boxes and two quarter-boxes. The arms overhead position equates to 7 hours per week, or 1¾ boxes.

Of course, you might not be able to be exact when it comes to sleep—there's some guesswork involved. Maybe you always fall asleep on your right side but wake up to find yourself in other positions. Maybe you have a sleeping partner who has observed you. Estimates are fine here.

SAMPLE 2: SLEEP MICRONUTRIENT TRACKER

2 hours per night side sleeping = 14 hours per week = 3½ boxes	ACTIVITY	Seated/Hip Flexion	Active Feet	Fixed Spine	Dynamic Arms	Forward Arms
	Side Sleeping (4 hours)	■		■		■
	Side Sleeping (4 hours)	■		■		■
	Side Sleeping (4 hours)	■		■		■
	Side Sleeping (2 hours)	▨			■	▨
1 hour per night with arms overhead = 7 hours per week = 1¾ boxes	Sleeping on back, knees bent (4 hours)	■		■		
	Sleeping on back, knees bent (4 hours)	■		■		
	Sleeping on back, knees bent (4 hours)	■		■		
	Sleeping on back, knees bent (2 hours)	▨		■		
	Sleeping on back, arms overhead (4 hours)				■	
	Sleeping on back, arms overhead (3 hours)				▨	
	Sleeping on stomach (4 hours)			■		
	Sleeping on stomach (4 hours)			■		
	Sleeping on stomach (4 hours)			■		
	Sleeping on stomach (2 hours)			▨		
	TOTAL OF FILLED BOXES	7	0	12.25	1.75	3.5

🛌 SLEEP: WEEKLY MICRONUTRIENT TRACKER

ACTIVITY

Each row = 4 hours per week
Use multiple rows for > 4 hours/week
Use quarter-boxes to represent
1, 2, or 3 hours

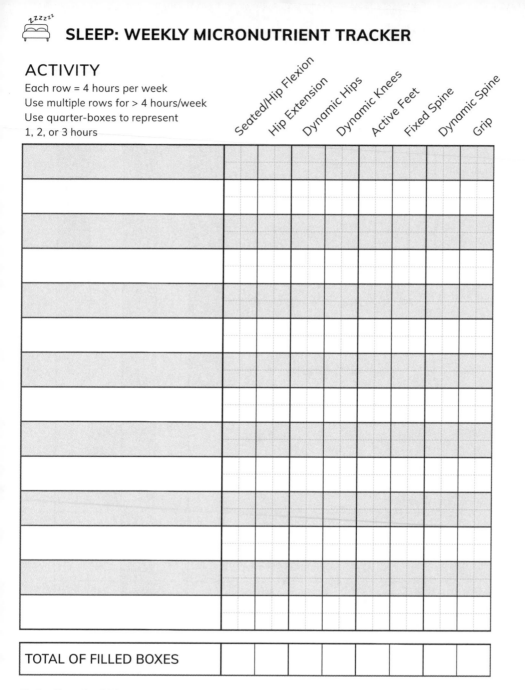

	Seated/Hip Flexion	Hip Extension	Dynamic Hips	Dynamic Knees	Active Feet	Fixed Spine	Dynamic Spine	Grip
TOTAL OF FILLED BOXES								

Circle all totals of 8 boxes or more ("Most": 32+ hours/week)
Draw a square around totals from 4–7 ("Moderate": 16–31 hours/week)
Draw a triangle around totals from 1–3 ("Minimal": 1–15 hours/week)

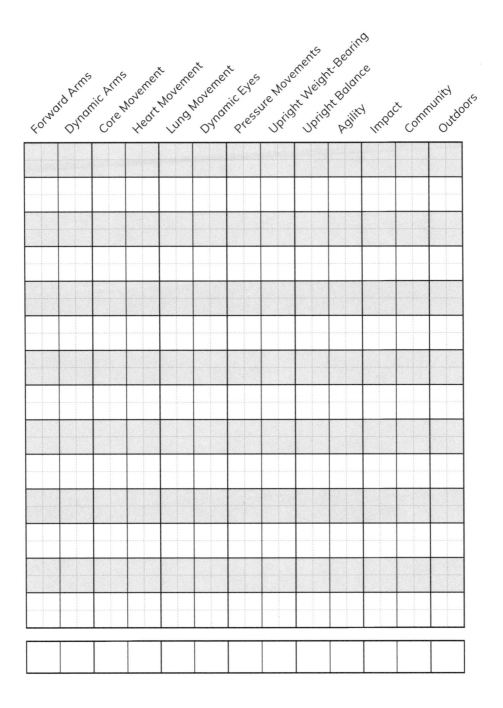

LEISURE: WEEKLY MICRONUTRIENT TRACKER

ACTIVITY

Each row = 4 hours per week
Use multiple rows for > 4 hours/week
Use quarter-boxes to represent
1, 2, or 3 hours

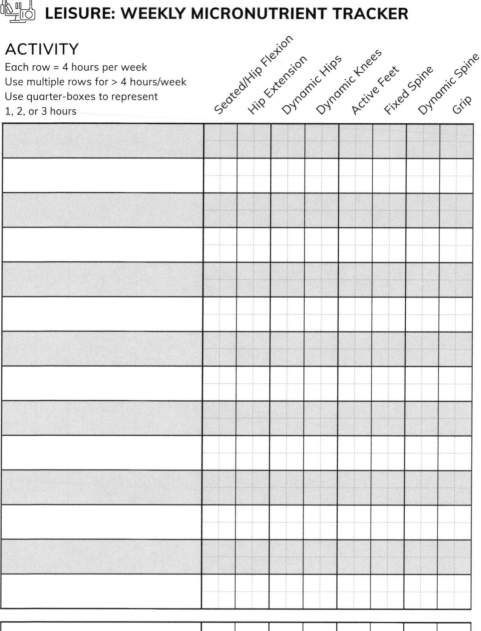

	Seated/Hip Flexion	Hip Extension	Dynamic Hips	Dynamic Knees	Active Feet	Fixed Spine	Dynamic Spine	Grip
TOTAL OF FILLED BOXES								

Circle all totals of 8 boxes or more ("Most": 32+ hours/week)
Draw a square around totals from 4–7 ("Moderate": 16–31 hours/week)
Draw a triangle around totals from 1–3 ("Minimal": 1–15 hours/week)

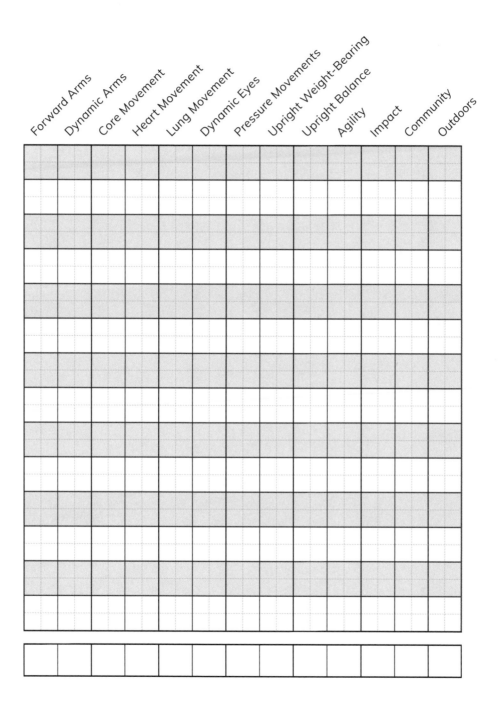

OCCUPATION: WEEKLY MICRONUTRIENT TRACKER

ACTIVITY

Each row = 4 hours per week
Use multiple rows for > 4 hours/week
Use quarter-boxes to represent
1, 2, or 3 hours

	Seated/Hip Flexion	Hip Extension	Dynamic Hips	Dynamic Knees	Active Feet	Fixed Spine	Dynamic Spine	Grip
TOTAL OF FILLED BOXES								

Circle all totals of 8 boxes or more ("Most": 32+ hours/week)
Draw a square around totals from 4–7 ("Moderate": 16–31 hours/week)
Draw a triangle around totals from 1–3 ("Minimal": 1–15 hours/week)

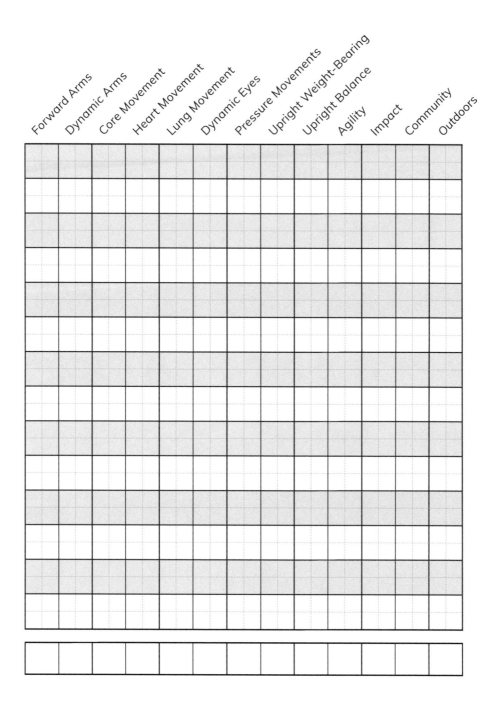

Forward Arms	Dynamic Arms	Core Movement	Heart Movement	Lung Movement	Dynamic Eyes	Pressure Movements	Upright Weight-Bearing	Upright Balance	Agility	Impact	Community	Outdoors

TRANSPORTATION: WEEKLY MICRONUTRIENT TRACKER

ACTIVITY

Each row = 4 hours per week
Use multiple rows for > 4 hours/week
Use quarter-boxes to represent
1, 2, or 3 hours

	Seated/Hip Flexion	Hip Extension	Dynamic Hips	Dynamic Knees	Active Feet	Fixed Spine	Dynamic Spine	Grip
TOTAL OF FILLED BOXES								

Circle all totals of 8 boxes or more ("Most": 32+ hours/week)

Draw a square around totals from 4–7 ("Moderate": 16–31 hours/week)

Draw a triangle around totals from 1–3 ("Minimal": 1–15 hours/week)

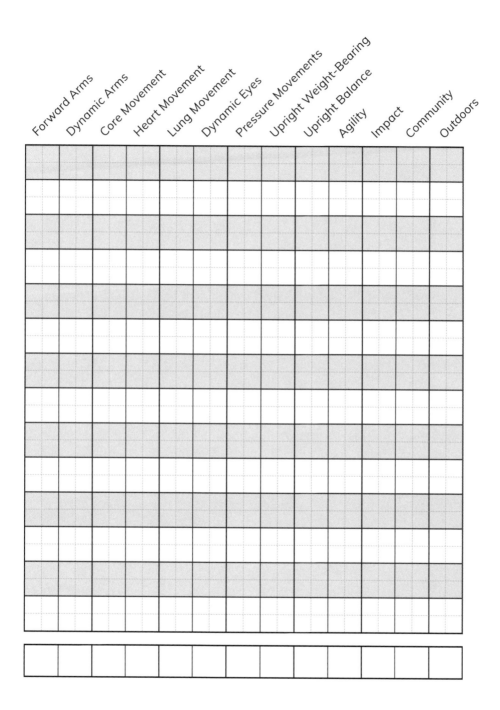

HOME: WEEKLY MICRONUTRIENT TRACKER

ACTIVITY

Each row = 4 hours per week
Use multiple rows for > 4 hours/week
Use quarter-boxes to represent
1, 2, or 3 hours

	Seated/Hip Flexion	Hip Extension	Dynamic Hips	Dynamic Knees	Active Feet	Fixed Spine	Dynamic Spine	Grip
TOTAL OF FILLED BOXES								

Circle all totals of 8 boxes or more ("Most": 32+ hours/week)
Draw a square around totals from 4–7 ("Moderate": 16–31 hours/week)
Draw a triangle around totals from 1–3 ("Minimal": 1–15 hours/week)

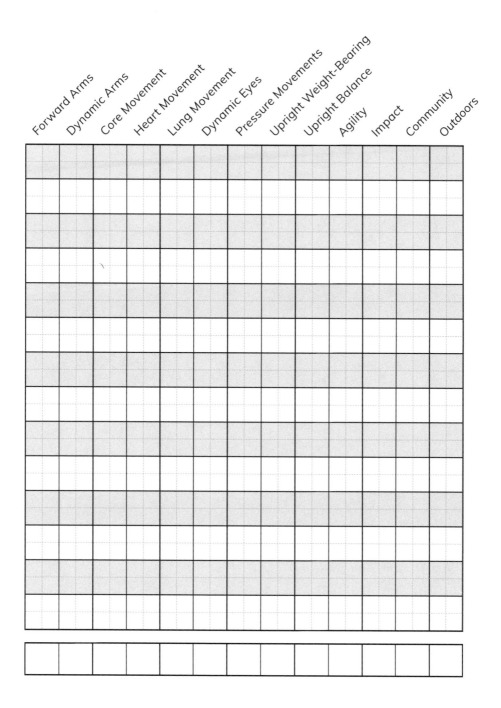

Forward Arms	Dynamic Arms	Core Movement	Heart Movement	Lung Movement	Dynamic Eyes	Pressure Movements	Upright Weight-Bearing	Upright Balance	Agility	Impact	Community	Outdoors

YOUR MICRONUTRIENT SUMMARY

Once you've filled out your activities and their shapes for each domain and totaled up the number of filled-in boxes at the bottom of each tracker, write the micronutrients into the appropriate section of the chart below.

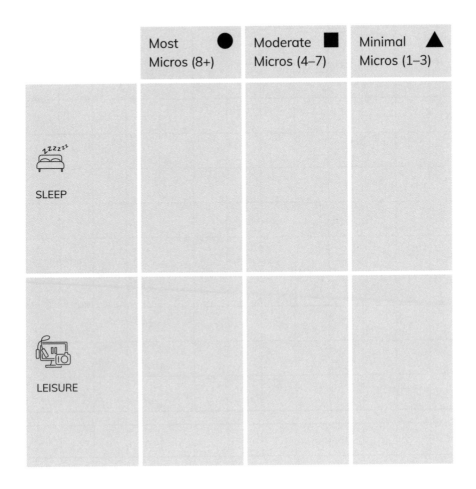

	Most ● Micros (8+)	Moderate ■ Micros (4–7)	Minimal ▲ Micros (1–3)
SLEEP			
LEISURE			

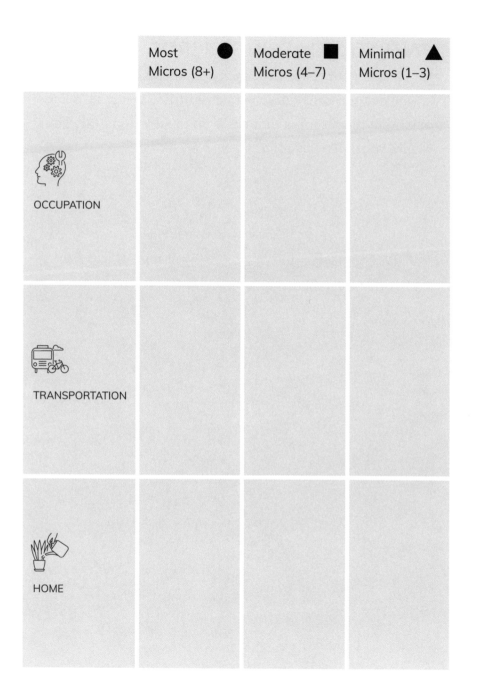

	Most Micros (8+) ●	Moderate Micros (4–7) ■	Minimal Micros (1–3) ▲
OCCUPATION			
TRANSPORTATION			
HOME			

CHART YOUR MISSING MICROS

1. Look at your tracker for each domain and in the appropriate column below list the micronutrients with a 0 score.
2. Circle the 0-scores that show up in ALL domains. Those are your Missing Micros.

SLEEP

LEISURE

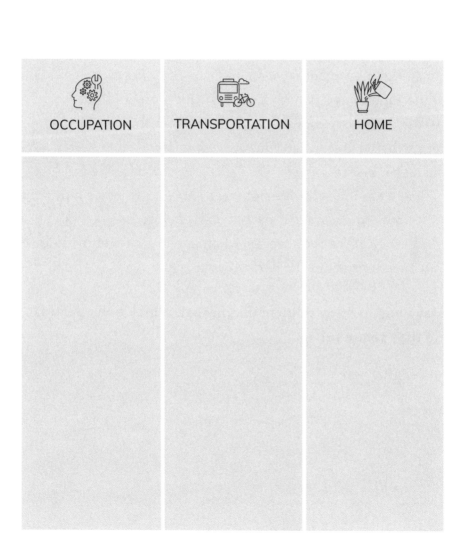

OCCUPATION

TRANSPORTATION

HOME

START CONSIDERING YOUR PARTS

I've put movement micronutrients toward the end of the book because the body's smaller movements are the hardest to assess. Some motions of the body are easy to spot, but others can get pretty technical. If you've ever gone to a physical therapist to rehab something, they've probably guided you not only through a specific set of exercises, but the particular form needed to place the movement *right where your body needs it*. It takes a while (and more training than just this workbook) to learn to assess your body movement to that degree, but you're on your way to being able to see your own movements more specifically than ever before.

There are times to think of your whole body as a single unit and times to consider it part-by-part. Before we start taking a closer look at the form we use when moving, answer the following questions about the parts of your body.

Have you had any clinical diagnoses? Which body part(s) do they relate to?

....

....

....

....

Do you take prescription medication or use medical devices (like a CPAP machine or braces)? Which body parts do these relate to?

....

....

If you ever take over-the-counter medication or supplements, which body parts do these relate to?

....

....

If you've had surgery, which body parts did it relate to?

....

....

Do any visits to the hospital, doctor, chiropractor, or any other allied health professional relate to a particular body part or area? Which?

....

....

Which, if any, body parts "alert" you on a regular or semi-regular basis?

....

....

List any body parts that hurt or aren't working to the best of their ability.

....

....

Do any of the health issues you worry about having in the future relate to a certain body part or area?

....

....

How do your "Most Micros" relate to the body parts/areas noted above?

....

....

How do your "Missing Micros" relate to the body parts/areas noted above?

....

....

Which body parts relate specifically to your Movement Why?

....

....

PLAN POINT

The body parts that keep me from moving in the functional and meaningful ways I would like to are:

....

....

Now that you have a more nuanced understanding of how your individual parts are currently being nourished, let's look at different ways you can start balancing them.

7

Moving the Rainbow

Every activity we do is facilitated by the collective shapes our body parts make. You've just identified the part-by-part movements you're getting a lot of, the ones you're not getting any of, and the ones in between. As you build your movement plan, you might need to balance your current micros out a bit: reduce those that might be interfering with your Movement Why and increase those that support it.

When you're running low on a particular dietary nutrient, you can usually buy what you need in supplement form. Vitamin tablets, mineral drinks, and protein and collagen powders are all dietary nutrients that have been isolated—taken out of their whole-food context—so you can add in exactly which nutrients you need, in the amounts you want.

Exercises are essentially *movement supplements*. Isolated from their practical or purposeful context, exercises are ways of bending, torquing, and smooshing different areas of the body to get the movement exactly where you want it to go: cardio exercise to move the heart and lungs, strength exercises for all different muscles of the body, and stretching exercises to target the joints you want to treat, etc.

SUPPLEMENT

Most of us live lifestyles that require at least some supplementing with exercise, especially when it comes to troubleshooting an injury or painful area. Corrective or therapeutic exercises are a way to inject movement directly into an underused area or to soothe the state of an overused area (e.g., releasing tension and relaxing a muscle that's used repetitively).

The following are examples of exercises and modes of exercise you can add to your day as needed. The list is certainly not exhaustive; you might have favorite exercises you can supplement with that aren't listed below. If you're unfamiliar with a move or exercise type, especially if you're trying to deal with an injured or painful body part, I recommend you find a teacher or therapist in person or instructional videos online. When it comes to micros, good instruction is very important. Look for qualified instructors who pay attention to good form in their cueing.

Hip Flexion

Plenty of seated activities flex and hold the hips at 90° (chair position). But hips need to move through their full range to stay healthy. To achieve this, supplement with deep squats, pulling your knees to your chest while lying on your back, standing leg lifts, deadlifts, cycling, elliptical, or rowing.

Hip Extension

Lunge stretches, lunge strength exercises, pelvic/hip thrusts, yoga pose Warrior 1, bridges, standing hip extension exercises, donkey kicks, CrossFit exercise GHD Hip Extension, walking, running.

Dynamic Hips

Deadlifts, side-lying leg lifts, the Pilates exercise Leg Circle, Zumba, hula hooping, dance, hip abduction/adduction exercises, the clamshell, lateral lunges, sumo/plié squats, martial arts.

Dynamic Knees

Jumping exercises (see "Impact" below), squats, lunges, Zumba, dance, step aerobics, step-ups, speed-skaters, roller blading, stair walking/running, goblet squats, martial arts, cross-legged sitting, the Pilates exercise Mountain Climber; also the yoga poses Eagle and all the Warriors.

Active Feet

Dance, calf raises, ankle point/flex, walking on cobblestone mat, toe lifts and toe spreading, barefoot walking or running (doing any exercise barefoot!), martial arts, slackline walking, Nutritious Movement exercises Calf Stretch and Top of the Foot Stretch.

Fixed Spine

Plank, the quadruped exercise Bird Dog, and any exercise that aims to "stabilize your spine."

Dynamic Spine

Spinal twists; yoga moves like Cat/Cow, Cobra, and Sun Salutations; Pilates Roll Ups and Roll Downs; Turkish get-ups; golf; throwing drills; paddling (kayaking, canoeing, stand up, etc.); martial arts.

Grip Movement

Kettlebell swings, farmer's carry, any barbell or dumbbell exercise, rock climbing, cycling, grip exerciser, rope climbing, TRX strap exercise.

Forward Arms

Cycling (road, mountain, spin), VR exercise games, boxing.

Dynamic Arms

Curls, chest fly or press with dumbbells, pullups, chin-ups, pushups, aerial silk, kettlebell halo, bent over rows, overhead press, rock climbing, TRX exercises.

Core Movement

Planks, side planks, leg lifts, crunches, rock climbing, gymnastics, rolling like a ball, the yoga Boat pose; also the Pilates exercises Hundreds, Roll Up, Teaser, and Criss-Cross.

Heart Movement

Stair climber, elliptical, running, aerobics, hills, weight lifting.

Lung Movement

Stair climber, elliptical, running, aerobics, hills, weight lifting.

Dynamic Eyes

Far-gaze exercises at varied distances.

Pressure Movements

Foam rolling, self-massage, fascia therapy balls (e.g., Tune Up Fitness).

Upright Weight-Bearing

Walking, running, sports, step aerobics, Zumba, stand up paddleboarding, boxing.

Upright Balance

BOSU standing, slackline, balance beam, the Nutritious Movement exercise Pelvic List, single-leg squat, stand up paddleboarding, martial arts; also the yoga poses Tree Pose, Eagle, Warrior 3, Dancer.

Agility

Lateral jumping, grapevine (as a dance move or fast, as in basketball warm-ups), pickleball or tennis (or hitting tennis ball against a wall), shuttle run, cone drills.

Impact

Jump training, arrested jumping, hanging drops, jump ups, jump rope, plyometrics.

Community

Ultimate Frisbee, team sports, walking/running with friends, family soccer. If there's exercise you love to do, see if friends or family will do it with you, which will increase the nutrient density of this period of time.

Outdoors/Nature

Hiking, kayaking, running, walking, squats in the backyard, stretching on the patio, Pilates on the beach…simply taking an exercise, any exercise, outside will get you more movement of your tiny environment-sensing parts.

SWAP

Another way people balance micros in a food diet is to change some of the foods they regularly put on their plate. There's a movement equivalent for that too: **swapping activities**.

If through filling out this workbook you've determined you need more Vitamin Upright Weight-Bearing, and you always use the stationary bike at the gym, you can swap it for something else at the gym—a workout on the elliptical or treadmill, or a game of basketball (which will *also* get you impact, dynamic arms, agility, and a dose of community). You can run a close errand on foot instead of driving (for more upright, hip extension, and outside time). You could take the grandkids to hit tennis balls against a wall or toss a Frisbee (grip, dynamic arms, outside time) instead of putting on a video (more inside, up-close eyes, hip flexion).

For community and nutrient micros, just add people you connect with or simply create an outdoor version and you've got your swap. Do you have a lunch break at work, during which you typically eat alone at your desk? Find a buddy who's also looking to feel better in their body and take your sandwich to go as you walk outside around your workplace. If you need mental downtime, skip the friend and take the walk with just your sandwich or with some of your favorite music (dynamic rest!). No matter what you're eating, lunchtime becomes a super-food in terms of how it's nourishing you.

SWITCH SHAPE

Another option is to switch your shape—do the activities you're already doing but with a different constellation. We've already covered replacing inactive sitting time with active sitting (for dynamic hips, knees, and core) in Chapter 1. Standing or sitting on the floor cross-legged at a low table for computer work, watching media in the evening while doing stretches on the floor, or changing the way you carry your backpack or purse (more on this below) are all ways to diversify your micros without adding exercise time or changing your activities. In case you were wondering, the food equivalent here is taking a staple or favorite food and switching the ingredients. Almond flour for wheat flour if you're reducing gluten or adding protein. Reducing the sugar or adding extra veggies (think sneaking shredded beets or zucchini into brownies) to a dish gives you a more nutrient-dense food that is similar enough to what you were going to make anyhow.

Have you ever heard "eat the rainbow" advice when it comes to dietary nutrients? *Carotenoids*—the elements of food that give it color—are another type of micronutrient our body needs. Each color has its own separate benefits, just as different body shapes have theirs. Instead of eating the same vegetables and fruits all the time, you add or swap fruits and veggies you eat often with different colored produce to make sure you're getting a little of everything you need.

Similarly, you can "*move* the rainbow". Instead of working (or walking, getting groceries, or playing a board game) the same way all the time, you change *the shape of the task* for another that gets you more of what you're missing and less of what you've got an excess of.

Below are common tasks that most of us do daily: sit, stand, walk, carry, and sleep. Making slight shape-changes to these activities is a simple way of getting more body parts the movement they need without having to add extra exercises.

SITTING THE RAINBOW

Active sitting is a great way to balance too much 90° hip flexion with more Dynamic Knee and Dynamic Hip time. It's also a way to get Vitamin Pressure-Deforming Movement and Core Movement. Active resting positions can take many forms, and mixing up the way you sit is a simple way to get not only more movement but also a wider variety of micronutrients. Look at the resting positions used around the world. Give each shape a try to see how each moves your body.

Which three positions pictured here are the easiest for you to get into? Describe.

....

....

Which three are the hardest? Describe.

....

....

Do you have a need for more Dynamic Hips, Dynamic Knees, or Core Movement? Which of these positions could you switch between or swap some of your chair-sitting for?

....

....

STANDING THE RAINBOW

Upright movement is a micronutrient because of the way standing loads both the skeleton and the muscles that hold it up. It takes a lot of work and endurance to stand, and the more we sit, the harder standing becomes. Remember the Law of Specificity? If we want to stand up well in the future, we have to practice this movement regularly. However, not all upright postures are equal; your alignment matters a great deal to how bones and muscles are loaded. Upright standing is really a category of positions that seem similar, but in fact subtle variations to the small curves of the spine, bends to the knees, and position of the pelvis all create very different loads. Technically speaking, standing at the bus stop with your body and head curled forward looking down at the phone is still an upright position, but it's not one that promotes better standing (and bone strength and balance) over time.

If you want to stay strong enough to stand well and keep your bones strong enough to support your body weight you need to fully and evenly load your bones with your body weight. You need to stand as *vertically upright* as you can. When the head or spine drop forward, their full weight is no longer borne by the bones below. Similarly, if the hips are shifted forward or slightly flexed, or the knees are slightly bent, there's less weight placed on the hip joints (a common site of decreased bone density and related hip fractures). This can happen on one side more than the other because standing with more weight in your right leg means there's less weight on the muscles and bones of the left hip. If you always stand that way, you end up with a side that's not strong enough to carry you well, which then keeps you standing—and walking—slightly favoring one side, and the weakness progresses over time. Studies show that people who stand and walk asymmetrically can also have asymmetric bone density. Balanced standing is the foundational step to stronger bones.

There's nothing wrong with any standing position, but each results in different outcomes, and your particular chronic/repetitive positioning is a set of loads that your body is adapting to. The micros in your "standing diet" affect how your body moves and feels every day.

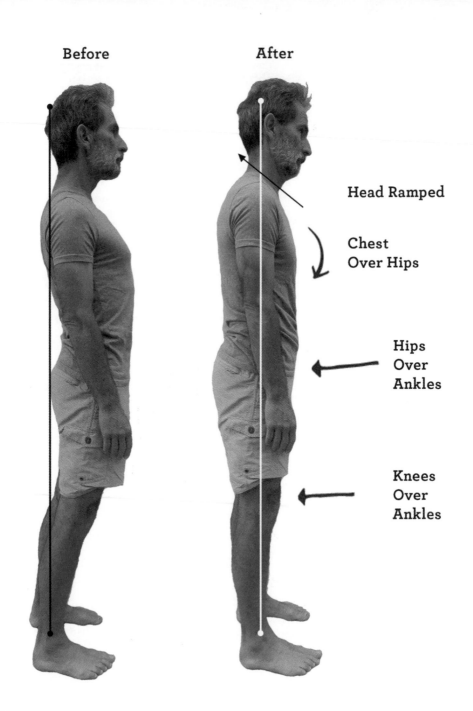

Before

After

Head Ramped

Chest
Over Hips

Hips
Over
Ankles

Knees
Over
Ankles

Micros Challenge

Start standing in a balanced way. We default to our most comfortable standing position over and over again, but with awareness and the help of a quick body check, you can adjust your body to maximize its muscular work and bone-loading. Every time you find yourself in your regular standing position, give yourself a quick makeover:

- » weight evenly balanced between your right and left foot
- » weight back off the toes and over the heels
- » hips back to stack the hip, knee, and ankle joints
- » spine tall.

Adjusting your stance just takes a few seconds, but if you can catch yourself multiple times a day, those minutes of balanced standing add up. You'll end up using body parts that aren't used to showing up when you stand up, which means the parts that have been doing more than their fair share of holding you upright can take a rest.

WALK THE RAINBOW

While all upright locomotion has a general shape—torso stacked over hips, arms and legs swinging back and forth—the nuanced shape the body makes when it's walking depends on your posture, the terrain you're walking upon, and the shoes you're walking in. You've just read about adjusting your posture as you can. Another thing you can do is mix up your walking surfaces. Walking on flat and level ground, climbing hills, stepping on

SWAP YOUR SHOES FOR DIFFERENT MICROS

Footwear affects your standing and walking micros. Conventional shoes are often elevated at the heels (yes, even athletic footwear). These cast your ankle in a slightly downhill position—the higher the heel, the steeper the hill your ankles are walking down, even when the ground beneath your feet is flat and level. If your calves are tight all the time, they might be low on Vitamin Uphill. Walking up hills gives your calves and ankles some of this much-needed movement; you can supplement with a Calf Stretch exercise, and you can also transition to flat shoes to reduce Vitamin Downhill if you've been overdosing on it.

Slip-on shoes are also slip-off shoes. They require constant Vitamin Toe Gripping to keep them on your foot. There are exercise vitamins to stretch out tired and tense toes that have been gripping shoes, but also consider switching to shoes that fully attach to the foot, no toe gripping required.

Shoes with narrow toe-boxes push toes together, preventing Vitamin Toe Spreading, even when you're walking, running, and jumping! This restricts the movement that the front of the foot needs, although you can get some in supplement form with Dynamic

Feet exercises. Or you can wear shoes that allow this movement and for your toes to spread without needing to add extra corrective exercise.

Finally, the primary way the feet's numerous joints get their movement nutrition is by walking over natural terrain that bends the foot into different shapes. Stiff-soled shoes prevent many of these smaller movements. To get that dose of Vitamin Texture, look for shoes with thin and flexible soles that offer some protection from the environment while still allowing your foot to move in the ways that keep it strong. Conversely, when feet are too flexible in one area and are bending excessively in the same spot, over and over again with every step, or if they're injured, reducing movement with a stiffer sole can be beneficial while you heal and/or learn to strengthen the foot so it can stabilize itself.

lumpy natural terrain, and taking the stairs all create unique loads, and therefore provide unique nutrients. The more you mix up your walk, the more walking nutrients you'll have in your movement diet.

How far is your regular walk? How long does it take you?

....

Describe up to three regular walking routes:

Do they have hills? Are they on human-made or natural surfaces?

1.

....

2.

....

3.

....

How much of your total weekly walking is on a treadmill:

0–25%, 26–50%, 51–75%, 76–100%?

....

Do you currently experience any pain while walking? Where and in which walking scenarios, exactly? Does this prevent you from walking as much as you'd like to?

....

....

....

Get your current most-worn walking shoes and go through this list. Every "no" you answer is a change you can make for more movement with every step:

Is the heel the exact same height as the toe? Y N

Can you twist the sole side to side, like you're wringing out a cloth? Y N

Can you curl up the sole toe to heel, like you're wrapping sushi? Y N

If you stand barefoot on top of the shoe and spread your toes apart, is the toe box still wider than your toes? Y N

How do your Most Micros or Missing Micros relate to the walking part of your movement diet?

....

....

Micros Challenge

This challenge will last four weeks at least (though it's so much fun you might want to keep it going forever!). You'll choose one variable for each week and change that variable on one of your walks every day.

Week 1: Change the distance of one walk every day.

Week 2: Change the terrain of one walk every day.

Week 3: Change what time of day you walk.

Week 4: Change the pace you walk every day.

You can choose from different variables—company, the load you carry, whether you're listening to music or not. For example: Week 1, day 1, change the distance of your walk and bring along a friend. Week 1, day 2, change the distance of your walk and carry something a little heavy. Week 3, day 3, change the distance of your walk and take off the headphones. Etc., etc. Get it? Fun!

CARRYING THE RAINBOW

How do you wear your bag or purse? On your back or shoulder? Which shoulder most of the time? Which arm do you use to carry your grocery bags? Your kids? Human bodies have been shaped by eons of carrying stuff in a variety of ways, each way working different parts of the body. Learning to mix up your carrying shape (what I call "vary your carry") will help you get a wider range of micros and decrease issues that arise from too much repetitive carrying shape, without needing to add extra exercise.

Carrying a load adds work to the arms, core, and legs. When a load is balanced evenly right to left (think having a backpack on your back with straps over both shoulders), most of the weight falls on the spine and legs; your abdomen must contract lightly to offset the backward pull. Move the backpack to your front and now there's a forward pull, requiring more work along your spine to compensate.

If you take that same backpack and put it over the left shoulder, the right side of the abdomen has to work more to stay upright. Put it on your right shoulder, and the left side of the abdomen must work. Hold it with your left arm and the arm itself now must work along with the right side of your core, and vice versa when you hold it with the right arm. This is all to say, you can get creative and place the movement right where you want it by choosing your carry carefully.

In what position do you usually carry loads? Describe.

....

....

Which part(s) of your body generally stops you from carrying heavier loads or carrying loads for longer?

....

Micros Challenge

Carry Your Bag Differently: It's easy to load up the same part of your body over and over again, but to work more of you, switch it up by carrying your preferred bag (backpack, purse, tote, etc.) over one shoulder and then the other. Wear it on your front or carry it in your hands.

Change Your Bags Up: For a full week, don't repeat a carrying method. Use your usual bag on Monday and a different sized one on Tuesday. Switch it out for a backpack on Wednesday and a duffel bag on Thursday. Carry a tote in your hands on Friday, and wear your backpack on your front on Saturday. (Don't

GRIPPING THE RAINBOW

Does all your gripping look the same? Same tool, handles, bars, or dumbbell shape? Repetitive grip-shape is limiting because it sets you up to be strong in a single way. You can increase the weight you lift in that particular way (i.e., you can get stronger in that way of moving), but you aren't making more of your body strong. Pick an exercise, like a pullup, chin up, or hanging or swinging from a bar, and change the hand-terrain. Seek out different bar angles, bar girths, shapes, textures, and wobbliness of the thing you're gripping. Each of these variables creates different gripping loads. Sometimes you don't need a bunch of exercises; a handful of them done in slightly different ways can take you far.

Which parts of your body would stop you from hanging or hanging longer? Hand skin? Sore shoulders? Something else?

....

What kind of surface do you usually hang or swing on? List all. (Smooth metal bar, smooth wooden bar, textured metal bar, tree branch, other?)

....

....

Micros Challenge

If you've been diligently hanging on monkey bars, you'll notice a very specific callus pattern forming. Hanging on different textures will load your skin differently, and you'll begin to build more uniform calluses over your palms and fingers. Get into the woods and see what it feels like to hang on a birch tree, then a maple, then a pine (or whatever trees grow in your area). Clamber up a lichen-covered rock or a moss-covered tree. Make a conscious effort to touch, climb, carry, and hang off a wide variety of natural objects whenever you have the chance. Your hand skin will thank you!

get stuck if these details don't apply to you; briefcases can be changed out for backpacks and duffel bags, for example.)

Vary Your Carrying Grip: We often carry stuff by placing a bag on the entire skeleton. So take that bag off your shoulders and put it in your hands! This is an easy way to layer the Grip Movement nutrient into daily life—no exercise time required. Next time you have a bag with handles, try these carrying shapes: Use your fingers like a hook, and hold the handles that way; hold the gathered straps like a rope; grip your hands together beneath the bag in different positions; change your wrist position—you can dangle your bag from the bottom or the top of a gripping hand; and when you need a rest but not a total break, the bag can go over your shoulder while you're gripping it.

SLEEP THE RAINBOW

Okay, we're finally going to get into the details of sleep.

Flat on your back, curled into a ball on your side, on your stomach with one arm over your head—these are examples of positions we can find ourselves in for hours each night; they're the big shapes in which we might sleep. But there are other shape-makers in your sleep environment: your mattress and pillow.

The Hadza and people in many other cultures of the world have sleeping surfaces that are more minimal that many of you reading this would consider. From animal hides to thin floor mats to piles of vegetation, much of the modern world sleeps in environments that create a great volume (hours!) of Pressure-Deforming Movements. Firm surfaces don't just push on you; they can also require your body to stretch out more. Thick, squishy surfaces don't require your shape to change—they accommodate it by bending around your body. You might already be sleeping in different shapes, but do you sleep on a mattress so firm it causes enough flesh squishing that you need to change position fairly often to be comfortable, or on one that's so squishy you can lie immobile for hours? Is it firm enough that you can feel reasonably comfortable sleeping Hadza-style on the ground or on a thin camping mattress if you need to?

As for pillows, Hadza use slightly mounded earth or balled-up textiles, or no pillow at all (arm joints can fold nicely into head support, it turns out). Your likely much larger pillow is affecting the way your head and neck are positioned for almost a third of each day. Imagine standing against a wall with your pillow behind your head. How far forward does it push your head? Do you feel limited in the movements of your head and neck during the day?

As you tune in to your sleeping movements, you might see how big an effect they have on how you feel every day. While sleeping on a very firm surface with little or no pillow might not feel as good as a deep-tissue massage, it has a lot of the same outcomes for your body, and the cumulative effects of getting something like a massage all night every night has surprisingly excellent outcomes.

How do your different body parts feel when you first wake up?

....

How many pillows do you sleep with? How tall are they?

....

How many nights a month do you sleep somewhere other than your own bed?

....

Is your sleep surface very soft, slightly soft, slightly firm, or very firm?

....

Micro Challenges

Switch Your Sleep Surface: To get your body used to moving against firm surfaces, progress through these challenges over a month.

Spend fifteen minutes on the floor, going through all of your sleep shapes before bed (that's some Vitamin Pressure-Deforming Movements right there!).

Switch to sleeping on the other side of your bed.

Spend some nights sleeping in a guest/other bed.

Sleep on something firmer, like a firmer mattress, a futon, or a heavily cushioned floor bed you make yourself, and see how you do. Try gradually less support until you're practically camping.

Or just stick to rolling around on the floor for fifteen minutes every day, making sure to press all your parts into the ground, for now. This is a great way to tenderize stiff parts that are un-practiced in regular squishing! P.S. If you're too tender to push into the ground, lay down a couple of comforters until your tissues become used to this particular type of exercise.

Pillow Switch: If you have a big pillow that raises your head a lot, slowly transition to flatter and flatter pillows until you're able to sleep with just a folded up T-shirt or without anything at all. Pepper in a lot of neck and upper body/upper back strength and mobility exercises to avoid neck kinks. Don't rush things! Expect this transition to take at least a year.

PLAN POINT

Considering the way your body is currently feeling and working, and your Movement Why, is there a need to reduce your Most Micros from page 190–91?

....

What supplemental exercises can you use to decrease impact? For example, if you've got a lot of Forward Arm, add in exercises from the Dynamic Arms category.

....

....

....

Which activity shapes can you switch the shape of?

....

....

Which activities can you swap?

....

....

Considering your Missing Micros on page 192–93, what are exercises you can do to supplement?

....

....

Which activities can you swap to get those Missing Micros?

....

....

Which activity shapes can you switch?

....

....

Nuance Matters

The movement micronutrients as you've just assessed them are a great place to start seeing your part-by-part movement. When you're ready, you can get even more nuanced—without a lot of training in anatomy or movement mechanics.

If I had a micronutrient box for every angle each joint makes and all the different loads that can be placed on those shapes, this book would fill an entire library, so instead I'll just say this: I've kept micronutrient categories pretty simple for easy assessment, but there's really a range for each. Let's take Vitamin Hip Flexion, for example. Say your trackers are filled with this box, but all of it is the "chair sitting" constellation. The chair-sitting shape is different from the constellations used while cycling (you've got more movement at the joint as well as more muscle contracting to make the legs move), which is also different from the *deep* hip flexion required to squat your butt close to the earth and the muscular work needed to get you up from that low position. All of these check the "hip flexion" box, but these activities are all different from each other, and getting regular chair time doesn't nourish the hips like deep squats do.

We can run through this range of variables almost infinitely—there are almost as many constellations as there are stars in the sky, or rather cells in your body, which is trillions.

One thing about stargazing is that the more you do it, the better you become at identifying constellations—and the same is true of evaluating your own body's constellations. Shapes will start popping out at you, both in yourself and when watching others. At this point, you've done so much self-assessment you'll be much more able to recognize your own undermoved areas, and you'll be ready to explore the nuance of micronutrients. (For guidance on delving into these, see page 252.)

Now, to the movement plans!

8

YOUR
Perfect
Movement
Plan

At this point, you've done the work to figure out what your current movement diet looks like, including which macros or micros you need to balance. But just because you've figured out everything you need doesn't mean that "go get all of these moves" is your plan. That approach is likely too big and too boring to be effective. The best movement plan for you isn't one that addresses *every* single movement nutrient you need—it's a plan that increases the movement-nutrient density of your life in a way that you can manage. Your perfect movement plan is one you can stick to!

DATE YOUR PLAN

Life is constantly changing, and your plan should keep shifting too. Even if your Movement Why never changes, other things will, which is why I recommend creating a new plan every quarter or so. Updated plans don't have to be completely different, but any time you identify a new stage in life—injury, pregnancy, new job, new empty nest, and even season depending on where you live—you'll need a plan that works in the new context. There are no hard and fast rules when it comes to movement programming duration—there are just too many variables that affect outcomes to make rules—but twelve-week programs are often used in both fitness and movement-research realms. Following a program for twelve weeks gives you enough time to see how your body is adapting physically, and your mind enough time to adapt to new habits. Put the date on your plan sheet and save your sheets to see how your plan has shifted over time.

FIND YOUR FOCUS

Every movement plan needs a Focus, the physical change you're moving to create. All movement is beneficial in some way, but you're using the Law of Specificity here to pick movements that ultimately serve your Movement Why. This could be getting your body able to do something specific: cross the monkey bars, walk all the mileage necessary for your once-in-a-lifetime vacation, get up off the floor with ease, or reduce stress. It could also be a focus on the state of your tissues: rehabilitating your injury, working on strengthening bone or muscle, or increasing your cardiovascular endurance.

When selecting your plan's Focus, you'll need to consider your Movement Why, your missing macros and micros, and your age and stage of life. Even if your Movement Why isn't directly about fitness or performance (maybe you just want to be able to play your banjo until you're a hundred years old) it still depends on your physicality in some way. What shapes and strengths of your body will it take to keep you moving in a way you find functional and meaningful? There are so many activities and exercises to choose from. The more specific you can be when picking your focus, the easier it is to determine the best environmental modifications, dynamic activities, and exercises to get you there.

I'm a forty-eight-year-old perimenopausal woman, mother of two, working outside of the home in a computer-based job for the most part. My personal movement plan is generally focused on maintaining overall muscle mass (see *perimenopause*), and keeping my long-distance walking parts strong, mobile, and pain-free—with extra focus on my pelvis and hips (a problem-prone area for me due to an old surgery and scar tissue). I fit lots of upright movement (for walking strength) into my leisure, transportation, and home activities, and big-body work (for whole-body strength) into my leisure and home domains, and I use a handful of challenging-to-me active sitting shapes when computer-working and for leisurely media viewing to get me the micros I need for my hips.

In the throes of finalizing this workbook, I threw out my back (no worries, I'm better now). This back injury is one I'm familiar with; it shows up once a year, typically when I'm under a larger-than-usual amount of stress for an extended period of time. When I added stress-writing into the pressure-cooker of perimenopause (mine) and puberty (my kids', not mine), additional work and home projects, and the general heavy vibe of the world these days, something had to give: lumbar vertebra 4.

My regular plan went out the window with my back injury, but luckily I knew how to create a short-term rehab plan. A lot of my work was done lying down or in positions I could handle. I could walk laps in the pool, which was easier on my spine, and I bulked up the therapeutic exercises I've learned that help me restore the motion I immediately lose when my back goes out.

Looking back, I can see that I should have preemptively adjusted my movement plan to accommodate my increased writing activity and general life pressure. A session of more gentle body care or meditation to transition my body and mind back to calm, or some less-intense bouts of movement at the end of a long writing day when my nervous system was fried, would have been good additions to my plan.

This is all to remind you that our "perfect" movement plans are fluid. They change with our age and stage of life. You can have an overarching plan for where you are right now in life, but there are things that can come up—injury, illness (your own or a family member's), grief—and plans need to adjust. Despite what's happening in your life, **you rarely have to stop moving altogether.** When you understand all the options available, you can constantly customize your movement plan to your shifting needs.

Once you have your focus, you'll consider how you can make over the physical environment of each domain and select dynamic activities in each domain that fill in some of the movement-nutrient gaps you have discovered in your movement diet.

ENVIRONMENT MAKEOVERS

For each domain, identify the Environment Makeovers—places where you switch up the shape of your environment in some way. Makeovers take a little bit of leg work so you can, uh, get more leg work. These tasks could include altering things you already have (create a standing work desk by placing a box on your kitchen counter), acquiring something you need to facilitate the movements you want, or doing some research—find new walking routes, community gardens to join, or exercise instruction. I've listed some examples below. Don't get sidetracked if there's something you can't get right now; there are likely many changes you can make without a new bike or balance board or pullup bar. Remember, **the perfect plan is one you actually stick to!** Only add the changes you can handle to your plan.

The nice thing about environmental changes is, once you do the work to make them, these changes help create more movement over and over again, without you needing to do additional work. So, each time you revise your plan you can tackle just a couple of these environmental changes, if you'd like. Over time, you'll slowly build a sustainable dynamic environment that supports your body and your Movement Why every day.

Sleep

- » Switch mattress (firmer mattress, mattress on the floor)
- » Sleep on padding on the floor (futon, pad, etc.)
- » Thinner pillow

DRESSING FOR MOVEMENT

An easy environment to change is our clothing. We often have clothing we recognize as "good for exercise"—flexible and non-restrictive. Does that mean the rest of our closet is sedentary clothing? Getting dressed each day in jackets that prevent your arms from reaching overhead, pants that don't allow the knees and hips to articulate, belts that keep you from bending over at the waist, and shoes that freeze your feet and ankles into a repetitive shape can keep you from getting more of the movements you need. Totally reforming the dress code in your domains might prove a tough job, but searching for different cuts and materials that are more movement friendly is one way you can make over an environment you spend all day in: your outfit!

» No pillow

» Sleeping bag outside once a week

Leisure

» Set up a slackline

» Create a standing gaming station

» Get a bike (basketball hoop, gym membership, tennis racquet)

» Plan a new walking route

» Find a community garden to work in

Occupation

- » Wardrobe: get minimal footwear/ dynamic clothing
- » Standing/variable desk
- » Set up your workspace by a window/create an outside workspace
- » Move everyday items to a spot that requires you move to get/use them
- » Get a dynamic chair (or wobble board for standing balance while working, etc.)

Transportation

- » Get minimal footwear for walking/cycling
- » Get a bicycle/ skateboard/scooter
- » Add a post-it reminder to head ramp while driving

Home

- » Clear a dedicated space for movement
- » Minimal footwear (or "no-shoes in the house" rules)
- » Rearrange kitchen: plates/cups down low or up higher
- » Same for bathroom: toothbrush on top shelf, towels down low
- » Add a squat platform to the bathroom
- » Create a low-sitting area
- » Add low tables
- » Hand-powered tools (push-mower, hand beater, coffee grinder)
- » Create a hanging/ climbing station
- » Place exercise

equipment out to
prompt movement
bursts

» Set up outside space
for eating, games,
kitchen work

DYNAMIC ACTIVITIES

For each domain, list the activities you'll commit to for the duration of the plan. Be specific. List the number of times per week, or the length of time you'll be doing something. Again, there might be many activities you could add to each domain but add to your plan only the ones you can realistically commit to and do with regularity. Some domains might have more activities than others. See examples below.

Leisure

» Play tennis once a week
» Take Zumba (yoga, Pilates, kickboxing) once a week
» Follow online exercise class twice a week
» Weekly date/friend hike
» Do "daily exercises" while

watching evening show
» 7-mile bike ride
» Adjust head position whenever on phone/ media
» Knit outside
» Volunteer to unload boxes at the food bank

Occupation

» Take eye breaks by looking out a window at

something far away for a few minutes every hour,

and any time you're on a phone call

» Take walks around the workplace or outside
» Take phone calls standing or walking
» Do two minutes of stretching on the hour (set reminders!)
» Lunchtime walk
» Take the stairs quickly

(or repeatedly, on work breaks)

» "Mind your pelvis" when standing
» Adjust your posture (head ramp, stance)
» Use balance board when standing at desk or counter
» Use different grips and carries when lifting

Transportation

» Drive halfway to work and walk the rest
» Cycle partway to work and walk the rest
» Park as far away from your destination as you can

» Walk to the grocery store and carry groceries back
» Stand on the bus
» Head ramp while driving
» Take the stairs

Home

» Yardwork: use push mower and rake
» Watch TV rolling on floor (pressure-deformation)
» Carry basket instead of using cart when

grocery shopping
» Floor sit while folding laundry
» Eat meals on the floor or at a low table
» Walk the dog twice a day

- » Move dinner prep outside
- » Cycle to the grocery store
- » Mop by hand
- » Calf Stretch while washing dishes
- » Picnic-style dinner once a week
- » Stand and stretch/walk the field edge while watching kid's soccer game

DAILY EXERCISES

In your leisure domain, there is space to add a handful of daily exercises—exercise supplements targeted to your Focus—that you commit to do doing daily. These are your personal daily movement vitamins. In this section of your plan, you'll identify one to six exercises that address specific micros or macros you aren't able to work into your non-leisure time. The Dynamic Activity section of the leisure domain is also the place for any other physical training/exercise you want to add to your plan that doesn't fit into the rest of your day.

REQUIREMENTS MET

The last step to filling in your movement plan is checking the "Requirements Met" boxes at the top of your plan sheet. After reviewing the rest of your plan, check the boxes you've addressed. We can only do so much and make so many changes at once, so your perfect movement plan—remember, **one that you can actually execute**—may leave some boxes unchecked.

Save your movement plan sheets, and over time, you'll notice if certain boxes are always unchecked. In this case, your next plan can focus on addressing any long-unmet movement needs.

Moderate-to-Vigorous

Look through the dynamic activities listed in your domains. Does your plan get you the heart and lung movement you need?

Macros: Upright Locomotion, Carrying, Climbing/Clambering, Making Movements, Big-Body Work, Active Rest Positioning

Mark the boxes for each of the macros you've worked into your plan.

Missing Micros

Does your plan incorporate some of the missing micros you've identified?

Meaningful

Movement nutrients need to not only balance over our bodies, but also be in balance with our personal preferences, i.e., what we find meaningful and functional for our individual lives.

Dynamic Rest

Does your plan include movements that allow you to rest your mind and give your "attention brain" a break?

BUILD YOUR PLAN!

It's time! You're ready to build your plan! For inspiration, there are sample movement plans in the next section of the book.

Review them before starting if you'd like to see what a completed plan looks like first.

The questions I get most often about building a plan relate to the rigidity we perceive in all domains of our life. How can we be more physically dynamic when the containers we surround ourselves with are penning us in and pinning us down? I recommend you start thinking about the most malleable parts of your life first, then work down to the least malleable.

If you're fortunate enough to have non-workdays, vacations, and weekends, are you putting movement front and center when work or school isn't forcing you in a particular direction? Then look at before and after school and work time. Are you using your time wisely in these periods and prioritizing the movements you'd like? When you're "inside" the more rigid containers—at work, at school, at home—are you taking the movement liberties you can, choosing the most dynamic of the static options, spending any break minutes to your body's benefit? Can you adapt your home container, the one you control, to suit your movement needs better? And could you get suited up for movement each day, dressing in a way that allows the freedom for your arms to move overhead, legs to squat down, waist to bend and flex—so you're always dressed for movement success, and never have to avoid movement opportunities because of your clothes?

Sometimes the most rigid container of all is our own mind. When you look closely, you will see there are many levels you can start working on, and so many ways to get your body moving more that don't push up against anything but your own habits and ideas. Get flexible in your thoughts so your body can follow. Build the movement plan that works for you, not only for the exercise, but for a more meaningful and nourishing life overall.

Happy, joyful moving!

_____'s MOVEMENT PLAN Date:_____

My Focus:_____

Requirements met (mark all that apply):

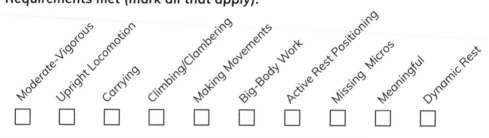

Moderate-Vigorous ☐ Upright Locomotion ☐ Carrying ☐ Climbing/Clambering ☐ Making Movements ☐ Big-Body Work ☐ Active Rest Positioning ☐ Missing Micros ☐ Meaningful ☐ Dynamic Rest ☐

SLEEP

Environment Makeover

LEISURE

Environment Makeover

Dynamic Activities

Daily Exercises

1.	4.
2.	5.
3.	6.

OCCUPATION

Environment Makeover

Dynamic Activities

TRANSPORTATION

Environment Makeover

Dynamic Activities

HOME

Environment Makeover

Dynamic Activities

Appendix:
Their Movement Plans

In the pages that follow you'll find sample movement plans for four people with very different movement diets.

Janet, 64. Restaurant server. Gets 3–5 miles' worth of steps at work four days a week, as well as some carrying.

Leisure: Daily nature walk of 3-4 miles for stress reduction, weekly yoga class and stretches at home for her lower back, weekly game of pickleball, occasional roller skating.

Home: Lives alone with minimal cleaning.

Transportation: Drives to work and weekend activities; walks when things are close.

Focus: Very cardiovascularly fit but has noticed reduction in muscle mass the last few years and is concerned with bone loss. Focusing her plan on getting more strength-building and bone-loading movement.

Xavier, 22. Student, almost exclusively sedentary: reading, studying, online lectures.

Leisure: Gaming (16–18 hours per week), plays ping-pong, has a gym membership but doesn't go often; lifts some free weights at home.

Home: Cooks occasionally, minimal chores.

Transportation: Takes bus to school, sometimes bikes.

Focus: Stress management and feeling more energetic.

Ruby, 51. Nurse, on her feet most of the workday doing a variety of tasks, including some lifting of patients, with some intermittent sitting; pretty exhausted after work.

Leisure: Tends toward low-activity leisure: reading, streaming TV, writing.

Home: Cooking, some gardening, spends weekends caring for aging father, running errands with him, helping him with day-to-day difficulties.

Transportation: Drives everywhere.

Focus: Is feeling her age and having some physical challenges at work—mostly just with energy; has a knee problem (Baker's cyst) that flares up with overuse; wants to do a walking vacation in Italy next year with her daughter. Focusing on walking stamina and pain-free knees.

Chris, 34. Massage therapist, with an average of twenty sessions a week. Stands the entire time work, plus lots of hand and arm movement and spine bending.

Leisure: Throws pottery a few hours a week. Does yoga, to try to restore his body.

Home: Has a three- and a six-year-old, cooking, cleaning. Avid vegetable gardener with a big garden—lots of digging, hoeing weeds, planting and transplanting, etc.

Transportation: Mostly drives, not a big walker or biker.

Focus: Starting to feel achier. Stays active through work and leisure (has never liked working out) but notices a decline in cardiovascular endurance. Wants better aerobic fitness and less lower back pain.

__Janet__'s MOVEMENT PLAN Date: Jan.–March

My Focus: Bone and Muscle Strength

Requirements met (mark all that apply):

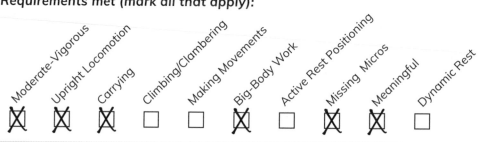

Moderate-Vigorous	Upright Locomotion	Carrying	Climbing/Clambering	Making Movements	Big-Body Work	Active Rest Positioning	Missing Micros	Meaningful	Dynamic Rest
☒	☒	☒	☐	☐	☒	☐	☒	☒	☐

SLEEP

Environment Makeover
» Trade mattress for firm futon

zZZzz²

LEISURE

Environment Makeover
» Get balance ball set up

Dynamic Activities
» Volunteer to stack boxes at Food Bank weekly
» 10-minute kettlebell routine Mon, Wed, Fri before work

» Roll out muscles with balls/foam roller while watching TV
» Online yoga class once a week
» Roller skate with grandkids twice a month

Daily Exercises

1. Foot stretches
2. Balance practice on BOSU
3. One pullup

4. Hamstring stretch
5. Pelvic List
6. 5 minutes jumping rope

OCCUPATION

Environment Makeover

» Buy minimal footwear with cushioning

Dynamic Activities

» 5-minute stretch break mid-shift

TRANSPORTATION

Environment Makeover

» Reminder note about ramping head in car

Dynamic Activities

» Head Ramp while driving
» Adjust whole body alignment when walking

HOME

Environment Makeover

Dynamic Activities

Xavier 's MOVEMENT PLAN Date: April

My Focus: Stress Management

Requirements met (mark all that apply):

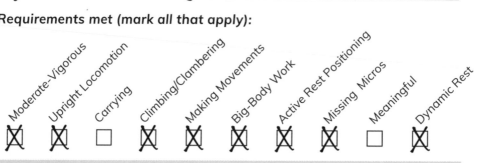

Moderate-Vigorous	Upright Locomotion	Carrying	Climbing/Clambering	Making Movements	Big-Body Work	Active Rest Positioning	Missing Micros	Meaningful	Dynamic Rest
☒	☒	☐	☒	☒	☒	☒	☒	☐	☒

SLEEP

Environment Makeover
» Sleep on the floor two nights a week (pressure)
» Ditch pillow

LEISURE

Environment Makeover
» Make standing gaming station with wobble board

Dynamic Activities
» Ping-pong with friends at least twice a week
» Lift weights in yard twice a week
» Use sauna at gym once a week

Daily Exercises

1. 20 minutes of outside walking

2. Lunge stretch

3. Upper back stretch

4. Dead hangs from pullup bar

5. Lie upside down over ball

6.

OCCUPATION

Environment Makeover
- » Get a ball chair for desk
- » Set up standing desk

Dynamic Activities
- » Set up a weekly outside group weekly study session
- » Stretch during online lectures (set a timer, at least 20 minutes)

TRANSPORTATION

Environment Makeover
- » Buy zero-drop sneakers
- » Get bike tuned for better gear shifting

Dynamic Activities
- » 5 minutes of breathing exercises on every bus ride
- » Stand on the way TO school when I take the bus
- » Walk to the library
- » Bike to the gym

HOME

Environment Makeover
- » Set up doorframe pullup bar

Dynamic Activities
- » Organize roommate group meals twice a week
- » Eat outside once a day
- » House clean on Saturdays with music

Ruby 's MOVEMENT PLAN Date: Oct.-??

My Focus: Walking stamina and pain-free knees

Requirements met (mark all that apply):

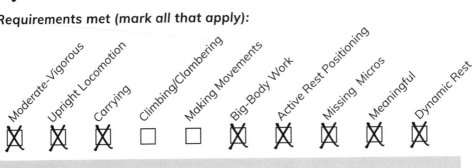

Moderate-Vigorous ☒ Upright Locomotion ☒ Carrying ☒ Climbing/Clambering ☐ Making Movements ☐ Big-Body Work ☒ Active Rest Positioning ☒ Missing Micros ☒ Meaningful ☒ Dynamic Rest ☒

SLEEP

Environment Makeover

LEISURE

Environment Makeover

Dynamic Activities
» Schedule 3-mile "coffee walk" in the park with friend twice a week
» Use Stairmaster at hospital gym for 15 minutes once a week
» Read outside in the hammock
» Watch TV shows stretching the floor 30 minutes a night

Daily Exercises

1. Calf Stretch

2. Hamstring Stretch

3. Quad Stretch

4. Calf Elevators

5. Pelvic List

6. Knee step-up/down

OCCUPATION

Environment Makeover
- » Use standing desk for computer work
- » Post reminder to Head Ramp at desk
- » Get minimal shoes with padding

Dynamic Activities
- » Focus on lifting alignment during patient transfers
- » Use 15-minute break for outside walk

TRANSPORTATION

Environment Makeover
- » Set a reminder on phone for commute times to remind about Head Ramp

Dynamic Activities
- » Park in the far corner and walk around to the far entrance of the hospital

HOME

Environment Makeover
- » Make space for floor sitting/lying by the TV
- » Set up nearby basket with massage balls and exercise bands
- » Add a pullup bar for short stretching and hanging sessions

Dynamic Activities
- » Eat one meal outside (non-work days)

Chris 's MOVEMENT PLAN Date: Nov.–Jan.

My Focus: Improve aerobic fitness and have less back pain

Requirements met (mark all that apply):

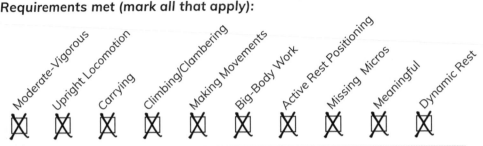

Moderate-Vigorous ☒ Upright Locomotion ☒ Carrying ☒ Climbing/Clambering ☒ Making Movements ☒ Big-Body Work ☒ Active Rest Positioning ☒ Missing Micros ☒ Meaningful ☒ Dynamic Rest ☒

SLEEP

Environment Makeover

» Switch to a thinner pillow

LEISURE

Environment Makeover

Dynamic Activities

» Pickleball Sunday mornings
» Family and friend hike once a week
» Yoga on Mondays

Daily Exercises

1. Finger/hand stretches

2. Chest stretches

3. Spinal twist

4. Hanging from a bar

5. Stretch out my body on/into the floor

6.

OCCUPATION

Environment Makeover

» Flat shoes/barefoot for working

Dynamic Activities

» 5-minute stretch routine (shoulders, chest, arms, fingers) between every client
» Adjust body alignment while massaging

TRANSPORTATION

Environment Makeover

Dynamic Activities

» Park 1.5 miles from work Tues/Thurs and walk
» Walk kids to library and back, carrying book load once a week

HOME

Environment Makeover

» Floor cushions for sitting

Dynamic Activities

» Watch posture while gardening daily
» Jump rope with kids 20 minutes a day

Move On: Micronutrient Deep Dive

This workbook was created to help you learn how to see the shape of your movement all day long—and why it matters. As I mentioned in Chapters 6 and 7, there are even more nuanced ways to see how your body is moving, and when you learn them, you can really precisely target the areas you need to.

You could go study anatomy and human movement to learn these nuances, or you can learn from someone else who's studied them. Good news: I love teaching anyone who wants to know how to spot areas of the body that aré under- or over-moving and how to adjust their alignment, daily movement habits, and exercises to get the movements they need specifically.

Movement classes: You can find over two hundred classes (with more added every month) full of nutritious movement micros in my Virtual Movement Studio. These classes range from introductory "snacks" to multi-class challenges, from five minutes to well over an hour. Because I'm a movement nerd, they're also full of information about your body and help you learn not just how to do the exercises, but why it helps to do them in certain ways. You can search classes by body part, macronutrient, or by special topic.

Find the Virtual Movement Studio as well as other movement programs at nutritiousmovement.com.

Movement books: Almost all my books are bursting with movement micros. Here are some to get you started.

Move Your DNA: Contains a lot of movement and load theory, plus digs into macros as well as some micros.

Rethink Your Position: If you want a book of movement micronutrients organized by body area, this is your book! Learn a lot more of the nuance when it comes to part-by-part movement.

Dynamic Aging: Want to focus your movement plan on aging well? This book is full of exercises and activities of daily living that address movements that relate to self-efficacy.

Diastasis Recti: If you identified an issue with your core OR upper body and want to work on the micros of this area, check out this book.

Whole Body Barefoot: Full of lower leg and foot micros! If you need more active feet, this book is the guide to all things lower leg and functional footwear.

Index

sleeping surfaces, 168,
174, 220
squatting, 35, 44–45, 110,
113–114, 122–124, 200–204,
225, 238
squat toilet, 114, 124, 233
sticky spots, 156–158
stiff/underused parts, 75, 78,
94, 157, 162, 166, 200, 222
supplementing with exercise,
149–150, 164, 169, 199–204,
212, 223–224, 236
swapping activities, 44, 59–62,
204–205, 208, 212–213, 224
switching shapes, 205–236

T

texture
hanging, 218–219
walking terrain, 161,
168–169, 172, 211–215
therapeutic exercise, 75, 76,
200, 230. *See also* physical
therapy
toilet. *See* squat toilet

transportation. *See* Domain of
Movement: Transportation.
Tune Up Fitness, 203

V

volume vs nutrient-density
of movement (table), 7–8.
See also movement-nutrient
density

W

Watson, Lyall, 79
weight-bearing, 88–91,
102, 112, 127. *See also*
Movement Micros: Upright
Weight-Bearing
while sitting, 35
while standing/walking,
111, 167–170, 209–211
weight lifting/carrying, 88,
102, 111, 123, 130, 133, 135,
162, 170, 216, 218, 242
willpower, 52–53
workstation (standing/floor),
36, 59, 114, 124

REFERENCES

To save paper and because you'll be accessing these sources on-line anyway, we've put chapter-by-chapter references and links for *My Perfect Movement Plan* online at nutritiousmovement.com/MPMPresources.

NEED MORE WORKBOOK PAGES?

Go to nutritiousmovement.com/MPMPresources *to find a downloadable PDF of the fill-in pages of this workbook.*

ABOUT THE AUTHOR

Bestselling author, speaker, and a leader of the Movement movement, biomechanist Katy Bowman, M.S. is changing the way we move and think about our need for movement. Her eleven books, including the groundbreaking *Move Your DNA*, have been translated into more than sixteen languages worldwide.

Bowman teaches movement globally and speaks about sedentarism and movement ecology to academic and scientific audiences such as the Ancestral Health Summit and the Institute for Human and Machine Cognition. Her work is regularly featured in diverse media such as the Today Show, CBC Radio One, the *Seattle Times*, NPR, the Joe Rogan Experience, and *Good Housekeeping*.

One of Maria Shriver's "Architects of Change" and an America Walks "Woman of the Walking Movement," Bowman consults on educational and living space design to encourage movement-rich habitats. She has worked with companies like Patagonia, Nike, and Google as well as a wide range of non-profits and other communities to create greater access to her "move more, move more body parts, move more for what you need" message.

Her movement education company, Nutritious Movement, is based in Washington State, where she lives with her family.